Do I Know You?

Do I Know You?

From Face Blindness
to Super Recognition

Sharrona Pearl

JOHNS HOPKINS UNIVERSITY PRESS   BALTIMORE

© 2023 Johns Hopkins University Press
All rights reserved. Published 2023
Printed in the United States of America on acid-free paper
9 8 7 6 5 4 3 2 1

Johns Hopkins University Press
2715 North Charles Street
Baltimore, Maryland 21218
www.press.jhu.edu

Cataloging-in-Publication Data is available from the Library of Congress.
A catalog record for this book is available from the British Library.

ISBN: 978-1-4214-4753-7 (paperback)
ISBN: 978-1-4214-4754-4 (ebook)

*Special discounts are available for bulk purchases of this book. For more information,
please contact Special Sales at specialsales@jh.edu.*

Contents

Acknowledgments

This is the best part of writing a book, and also, in some ways, the scariest. Despite keeping a running list in my head of everyone who has helped me produce this book, I am terrified of forgetting to recognize (groan) someone. We forget all too often that the academic world, and indeed (to be grandiose, and in this time of attack on the humanities I think it's worth being a bit grand) knowledge production itself, is collective. The temptation to claim novelty is strong, and the pressure is powerful. I get it. I've done it! "The first extended study of x"; "a new way of thinking about y"; "a never-before-seen instance of z." But we need to be careful of such formulations; too often we in the West ignore the work of others, particularly those with less power or positionality or recognition, or those who work in languages other than English, or in the Global South. And too often we ignore the rich and rewarding ways that all our work builds not only on what came before but around those with whom we are in community. Yes, I spent many hours alone in a room with a computer (and a pen; I'm a bit old-school). Yes, I read many things all by myself. Yes, I wrote the words in these pages independently. And *yes*, I formulated my ideas through countless conversations with friends and colleagues, the brilliant and generous people to whom this book, and my pride in it, is indebted. I am grateful for your insights, your feedback, your support, and your patience. I cannot begin to list the ways that being in community with you has made my work, and me, better. I am proud to know all of you and prouder than I can say to call you my friends.

Hannah Zeavin: I'm so grateful for the day you slid into my DMs. You have been a cheerleader, an interlocutor, and a heroic reader. I am so excited to keep thinking with you. Michael Gordin: you are a structure doctor extraordinaire. I wrote in my dissertation that you are as smart as you are nice and that everyone who knows you knows what a compliment that is. That is, astonishingly, even more true now than it was then. Paul Lerner: you generously responded to a call on Twitter to help me think through this book and ended up reframing my entire approach by putting me in touch with the

Wende Museum. Thank you for helping a once stranger with your characteristic generosity.

I have so many cheerleader colleagues who remind me on a regular basis that despite the challenges of academia, there is so much goodness, kindness, and selflessness. Alex Geisinger, Amelia Hoover-Green, Anil Kalhan, Scott Knowles, Ros Remer, Chloe Silverman, Amy Slaton: you welcomed me to Drexel with open arms and helped create a true community for me. Susan Feinstein, you have my back and make my life easier in so many ways, and I promise I notice. Jesse Ballenger, Merritt Brockman, Connie Perry, and Gina Yacovelli: I owe you all so much, and I am so grateful to you for modeling for me how to stand up for your beliefs even at personal risk. Aila Luneau provided key tech support for this book and so many other things. Nick Dames, Peter Decherney, Michael Delli Carpini, Nancy Hirschmann, John Jackson, E. Patrick Johnson, Amy Jordan, Marc Lafrance, Grace Lavery, Carolyn Marvin, Shannon Mattern, Erika Milam, John Durham Peters, Amit Pinchevski, Rabbi Danya Ruttenberg, Jonathan Sadowsky, Alyssa Sepinwall, Michael Socolow, Deb Thomas, and Barbie Zelizer: your ongoing support means the world. I can't tell you how much I appreciate your many great and small kindnesses that I have been lucky enough to receive probably far too often. Daniel Goldberg: I admire your integrity and your thoughtfulness, and I am so glad to know you. Nora Jones and Megan Voeller: I look forward to many more dinners and discussions, mostly not about bioethics and health humanities. Paul Saint-Amour: so many of the ideas here are better because I thought them with you. Riva Lehrer: working with you was one of the bright spots of the pandemic, and I am so proud of our collaborations. I learn so much every time we speak, and I look forward to many more conversations. Claire Raab: you are a superhero, and I am in awe of you every day. Thank you doesn't even begin to cover it.

This book in many ways began because Matt Price and Becky Anhang Price sent me an article; it is one of many beginnings we have shared, and I look forward to many more. They inspired me to reach out to Josh Davis, who became my "super recognizer researcher on call," answering questions, agreeing to be interviewed, and offering tremendous thoughtfulness and insight. I am also grateful to Richard Russell and Jennifer Jarett, who generously gave their time to speak with me about their own expertise in super recognition. Any errors are of course my own.

Joes Segal, Marieke Dros, and Kathryn Ung at the Wende Museum were enormously helpful and supportive, and the chapter on Peter Bochmann would simply not have been possible without them. Go visit the Wende—it's worth it! Elizabeth Neswald generously translated the archival material from the Wende. Parts of chapter 3 and chapter 8 appeared in an earlier form in the open-access journals *Semiotic Review* and *Disability Studies Quarterly*, and I am grateful for permission to reuse them here. Mike Burton kindly gave me permission to use the Glasgow Face Matching Test, and Jaipreet Virdi allowed me to use her photo of Checkpoint Charlie, both in chapter 5.

This book has benefited enormously from feedback from audiences at the Canadian Society for the History of Medicine; the Surveillance Studies Network; Archival Kismet; the Society for the History of Technology; the History of Science Society; the Health Humanities Consortium, Drexel University STS Works-in-Progress Seminar; the Kline Law School Colloquium, the Drexel History Department brown-bag seminar; Tulane; the Gallatin School at New York University; the Visual Studies Program at the California Institute of Technology; and the University of Texas Medical Branch. I am also grateful for the thoughtful and useful feedback from the three anonymish (it's a small field!) reviewers.

Matthew McAdam is a true editor; he has offered support every step of the way, and the book is considerably better because of his detailed, penetrating, and careful edits. I hope everyone is lucky enough to work with an editor like Matt. Adriahna Conway helped usher the manuscript into its final form, as did David Goehring's and Charles Dibble's generous copyedits and the excellent indexing skills of June Sawyers. Steven Fraser provided beautiful cover art, and I feel honored to have been able to work with such a talented face-blind artist.

When John Donne wrote that no man is an island, he didn't know that islands are good, actually—when they are populated by ladies. To Courtney Thompson, Janet Golden, Kelly O'Donnell, Aparna Nair, Elizabeth Neswald, Lauren MacIvor Thompson, Amanda Mahoney, Kylie Smith, Cornelia Lambert, Sarah Swedberg, Jai Virdi: not only have you made this book better (and, maybe, enabled it to exist at all); you have made me better. Thank you for the generosity, for your endless encouragement, for the sharp edits and critiques, for the suggestions about structure, for the opportunity to think aloud and improve (and discard) my arguments and ideas, for the laughter, and for the love. For so, so much love; I am not sure why I have been lucky enough to be among you, but I am honored to be in your community and humbled by your brilliance and kindness.

My mother, Susan, continues to be a model for lifelong learning and passion for open-minded discussion, debate, and engagement. My brother, Mayeer, and his wonderful wife and children always remind me how important—and how much fun—family is. And to my own family: you are, simply, everything. Aria, Melilla, and Yishai: words cannot express how proud I am to be your mother. Your kindness, your compassion, and your care for one another and the world amaze me every day. The world is a better place because you are in it, and I admire how deeply you understand what a responsibility that is. You give me hope. Ben, you inspire me, you support me, and you believe in me. I am so grateful for the opportunity to follow our dreams together in this beautiful life we have made.

It is written in *Pirkei Avot, Ethics of Our Ancestors* 2:1: עַיִן ,דַּע מַה לְמַעְלָה מִמְּךָ רוֹאָה וְאֹזֶן שׁוֹמַעַת, וְכָל מַעֲשֶׂיךָ בַּסֵּפֶר נִכְתָּבִין: דַּע מַה לְמַעְלָה מִמְּךָ, עַיִן רוֹאָה וְאֹזֶן שׁוֹמַעַת, וְכָל מַעֲשֶׂיךָ בַּסֵּפֶר נִכְתָּבִין ("Know what there is above you: an eye that sees, an ear that hears, and all your deeds are written in a book"). As I have written here, there are many ways of seeing. There are, equally, many ways of hearing. And there are many deeds in this book: I hope you enjoy it.

Do I Know You?

Introduction

Inventing a Spectrum

This book is a way to build a history—of relationships, of health care, of diagnosis, of disability, of storytelling and listening, of technology and computing, of racial capitalism and surveillance, of tacit knowledge and labeling—through faces. It's part of my larger agenda to craft a cultural history of the face.[1] I think about the kinds of knowability that come through taxonomy and categories, what is gained and what is lost with labels. I examine the interplay between care and power in what counts as disability and what counts as skill. I look at the invention of difference through the face and consider how cases create types, and how the individual and local interact with the general and global. I do this through the particulars around face recognition, and, importantly, through a reflection of my own role in collecting these particulars to make one among many possible sets of general claims. I use the face as my guide to explore how medicine, science, media, and sociality converge to create a category and put a name to lived tacit knowledge. To do so, I turn to the knowability of people through their faces by learning about specific people and how they recognize faces. Or don't.

In this book we'll meet people on both extremes of the face recognition spectrum, and we'll learn about the ways that their experiences are hard. We'll learn too about some of the people who study them and the work they have done to establish parameters, define a category and a research agenda, and perhaps leverage this new field of study in professional, practical, and monetary ways.[2] Before we encounter them in this book, some facts will make it easier to understand what's happening when most people see faces and when people on the extremes see faces. These facts will also make it easier to understand what most of us can't really understand: what it's like to not recognize a face. Meeting them will be part of the point: my own positionality as a historian and scholar, in a particular city, with access to certain networks, helps create the story. I write here about the creation of a category—face recognition—and in so doing, I help create it and I also contribute to the ongoing narratives around it.[3] The extremes of face recognition are occupied by relatively small numbers of people, but they create a spectrum that encompasses an entire world. We will consider the pivot from local to global stories, looking at how one person's lived experience of face recognition is both theirs alone and also, in ways both soothing and disconcerting, can be shared by others. That's one part of the pivot between specific and general; another is through those creating the categories institutionally: people on the extremes of face recognition but also the people who study them, write about them, and build careers upon them. That group includes the scientists, and, by my writing about the subject, also me.

In broad strokes, this book is a biography of the face recognition spectrum and how it was established. I am interested in what I call here "spectrum thinking": the way that spectrums can become sticky at the extremes with labels that adhere strongly and are hard to adjust or shift, labels that ignore the nuances of what lies in between. In, say, gender and sexuality studies, the spectrum has become a language of fluidity, mobility, transition, and ongoing change. There is also a way that spectrums entail a kind of fixedness, particularly at their poles. That's not a critique: while some people comfortably inhabit fluidity, others quite happily exist and remain situated on the margins. But the notion of the spectrum itself sometimes—in the case of face recognition, for example—means that we don't think of the extremes together, despite what they may have in common. There's a loss in the adherence to categories of what might be "bad": face blindness; and what might be "good": super recognition. And while one carries with it significantly more challenges than the other, they're not so different. "Extremeness" itself links those on either end of a spectrum; in some cases, we ought to think of the extremes together instead of in opposition. People at both ends of the face recognition spectrum might share patterns of sociality; they certainly share some bare facts about how they know faces. For face-blind people and super recognizers alike, the ability to recall faces is broadly unconnected to their encounters with others. The relationships that the most extremely face blind and the most extreme super recognizers have with other people are meaningless as an index to recognition: the face blind remain face blind equally to all, including dear friends and family, and the super recognizers remember everyone equally.

Face recognition is deeply intertwined with practices of surveillance.[4] There is a meaningful difference between recognizing the people with whom we interact and being recognized for the purposes of tracking who we are and what we do—where we go, what we buy, and what we like. But the skill of face recognition is deployed in service of surveillance. Biometric tracking is embedded in a long history of sorting people; more recent deployments of face recognition technology sit in this tradition of using our bodies not just to know who and where we are but also to gauge our value to governments, corporations, police forces, and border control. I place face recognition technology within that same history of sorting people, showing the ways the face itself is part of the invention and assignment of difference. The story of biometric surveillance is also associated with the history of computing, but I take a different approach to these narratives. I am interested in the tradition of determining value through faces and spectrums and considering the ways that face recognition technologies have always been mixed with surveillance and control. I situate the personal relational and emotional stakes at the extremes of this spectrum, but face recognition is not just an interpersonal experience. There are broad technological and sociocultural stakes to the leveraging of face recognition, and especially super recognition, in a global economic and political context. While I focus largely on the extremes of face recognition, I also look at attempts to move to super recognition, both analogue, in the training of border guards, and digital, in the face recognition software that is now ubiquitous but was once an experiment in trying to mimic and thus learn about the human brain.

I also offer a plea for the health humanities. We must listen to patients. We must be attentive to relationships and experiences and how the lived body plays a key role in how we inhabit the world. Face blindness was clinically named only in 1947, and it took another sixty or so years for it to be broadly recognized and studied. Super recognition was named in 2009, which is yesterday in scientific terms. As this book argues, taking pictures of brains tells us only so much. Pictures are great, but they aren't the whole story. We can turn to humanistic approaches—including history, visual studies, literature, and communication—to understand more about health and medicine, and to make patients' experiences, lives, and diagnoses better. We can and must use these tools to improve both the experience of health care and its outcomes, and to reflect on the ways these tools are necessarily intertwined. Health humanities isn't just for providers, and it's not just about finding creative ways to express experience and use health to engage with the arts. It's about drawing on the humanities to enhance our understanding of the body, of power dynamics around health and healing, of the nuances of history and how they are inherent in the health system today. It's about patients' lived experience. It's about the fundamental questions of what makes us human.

The paradox of face recognition is that it is the most tacit of knowledges. It's an ability that we don't realize is an ability, an ability that we don't realize we can lose, or that we maybe never had in the first place. It took so long to enter the documentary, medical, and rhetorical record because we didn't even have a category for it in the first place. Those who fell outside the experiential norm didn't have space to understand their own experiences as something beyond a lack or a particular knack until we began to really listen to what they were saying. Through the emerging discourse around face recognition, we can think about the construction of disability as a category and how ways of accessing the world become valued as a lack or a superpower. We can see the stakes for those designations in how these categories are studied and how they are described, and what that means for how they are experienced. In a way, this is an extended case study for the role of description and lived experience and for narrativizing the brain-body, not just for its own sake, but because we might actually learn how some processes work. The second half of this book charts precisely that process, with a theoretical investigation at the very end about how systems work, how they obscure their own workings until they break down, and how face recognition fits into that.

This is not a detective story. Nor is it an origin story, despite the deep attractions of an origin story. I am not going to delve deep into case histories of patients long past to track down the missed and hidden examples of face blindness. I'm not going to present marvelous magical stories of superhuman feats of identification and declare these figures to be unrecognized super recognizers. The current thinking about face recognition means that there were and are plenty of people who fall on either end of this spectrum who don't know, and don't even know that they don't know. There is something satisfying about conferring the relief of a label upon people who always felt different but were never able to articulate why.

Labels, categories, and diagnoses invented in one time do not make sense when mobilized retroactively to another. They are all a function of the particular context in which they emerge. They are flexible and emerging entities that can change going forward. But they should not be applied looking backward.

And yet. I admit I'm ambivalent: we know the risks, but there are definitely losses if we reject retrospective diagnoses out of hand. What is the project of history if not to excavate what once was and make sense of it in terms of what is? To put it another way: can we not imagine that there were once super recognizers who rose to prominence as queen's aides, politicians, merchants, or rulers? Is it not possible that there was once and always someone who did not recognize the face of others, and did not recognize that she did not recognize others, instead feeling forever as if those others had a superpower while she was just ordinary? Or less? What would be wrong with that? Should I track down those antique references and say, "Aha: here it was, and it was ever thus"? (Is this a detective story after all? What is invention if not an origin story?) We look and maybe even fetishize origins to make sense of the present, and the language and narratives of the past are very much a part of the constructions of the present. What is the project of history if not to situate change over time, starting with a particular moment? But that will work only partially here: while I'll offer the neurological history as a kind of origin story starting from the nineteenth century, face blindness as a category of experience really emerged only in the twenty-first. I have only snippets and single cases, but they too have deep meaning and value, and not just to those involved. Each case matters individually, but we can also think of cases together. The search for origins goes backward in time, but the narrative of invention goes far forward; we can't understand the approaches to face recognition without grounding developments in a long trajectory, but we can't make sense of it as a category without focusing on a short one.

This book is many things: a biography of a diagnosis with (at least) two beginnings; a deep look at the history and stakes for the face in human encounters; a way to think of binaries together; an examination of how local networks and individual experience become official history; an exposure of tacit knowledge through failures in the once-unnamed system of face recognition; and a grounded plea for the health humanities—but it is not an extended origin story. Beginnings, what we mark as beginnings, matter in how we record history; historians are, as I'll discuss, players in the stories we write. I'll excavate the cases that look like something we might recognize, if we squint, as being like face recognition. That's one way to start a book, and we'll do that here with a kind of origin in the nineteenth century. And then we'll try another way, turning to its invention in the twenty-first.

Face Recognition: A Primer

A lot of things have to happen in the brain for us to recognize faces. It's actually less extraordinary that some people can't do it than that so many of us can. People, in general, are pretty good at recognizing faces; a face may often seem familiar even when we can't remember names or context. For some people, face recognition just doesn't happen as well, and for others, it doesn't happen at all. And it's one of those things that seem to be impossible to understand how it works in others. Because I am a person who recognizes faces pretty well, face blindness just doesn't make sense to me. There are all kinds of metaphors and explanations and attempts: imagine you were shown a picture with a pile of Legos of different lengths and colors. Imagine the picture is then taken away. You would certainly know and recall that it contained Legos, and maybe even some broad features of color and shape. But you are unlikely to remember precisely the order and configuration of each piece. Perhaps that is what face blindness is like: people can remember that they saw a face, with eyes and a nose and a mouth. But *which* eyes, *what* nose, and *whose* mouth—those details disappear immediately after the face is gone. Voices, to those with an ear or who have developed this adaptation, may offer consistent clues. Gait is often ingrained and can be linked to a particular person. Hairstyle and shape, distinctive piercings and moles and tattoos and glasses all contain lasting resonance. Face-blind people rely heavily on such markers. But many of these markers can change, sometimes with no notice. That leaves face-blind people without reliable ways of recognizing others; a change of hairstyle or clothing may mean that someone who was identifiable in the morning becomes impossible to distinguish in the afternoon.

Or maybe it's more like seeing a line of cheetahs and being told that one is called Bob and the other is Shelley. You may be able to identify Bob and Shelley (and Amit and Raoul and Carol and Noa). You may not. Certainly, some people can, especially those who have worked with these cheetahs every day. To them, the distinctive markers that separate each cheetah from the other could not be more obvious. There's a lesson here about categories of familiarity, what we think is knowable and valuable to know, how individuation works and what fields come to bear on the process. We might expect a zoologist or anthropologist to recognize cheetahs. Not a historian. We may expect people to recognize the kinds of face with which we are most famil- iar. We even have acronyms for this "other-race effect" or "cross-race effect": the ORE/CRE. And that itself ramifies in surveillance, criminal justice, and practices of detection. Those who are more easily recognized are less likely to be falsely identified. Those who are doing the recognizing—those in power, white people—better recognize those who look like themselves.

Perhaps over time and certainly with some training, the difference be- tween cheetahs may become obvious to some—a patch here, a color there. But will one set of cheetah eyes ever look different from another? Probably not for me. Cheetah blindness is not face blindness, and this example exposes a profound bias in what counts as identifiable. There are, of course, other ways of thinking about faces and other ways of thinking about human faces. There are other ways of thinking about blindness and what it means not to see and not to know another; Saidiya Hartman has taught us about those obscured by history, photos in an archive without records or notations, faces in the past without recorded stories.[5] Face blindness is profoundly individ- ual, between one and another, between one and everyone. Historical blind- ness is about everyone to one, everyone to everyone who is unknown. But we can continue to look to the past, and we can continue to try to see.

For about 1–2 percent of the population, no amount of staring at a face will help. No amount of training will help. Color-blind people cannot be taught to see color. Face-blind people cannot be trained to recognize faces. Profoundly face-blind people simply will not recall the features of a face. They can see them—eyes, nose, mouth, and so on—but when they are not look- ing at them directly, they simply cannot recall particular features with any specificity. Others can do it slightly better and so on and so on, with the bulk of the population mostly able to mostly recognize most faces. Relationships help. Repeated exposure helps. For most people paying attention helps. Shar- ing a powerful moment helps. And, while memory and face recognition are broadly unrelated at the extremes, memory, taken in conjunction with every- thing else, helps. The top 1–2 percent of the population can recognize faces better than anyone else. And everyone else can't be trained to do that either. And how odd it would be to condemn a color-blind person for not being able to distinguish red and blue. And how odd it would be to castigate someone for failing to recognize a face, or, indeed, for recognizing only some faces rather than essentially all of them.

That's how spectrums work, by design. But, as we'll see, face recognition wasn't always a spectrum. When face blindness was first described, scientists thought that it was a pathology, a disability that some had to greater or lesser degree. And everyone else could just recognize faces more or less equally well. That changed dramatically when Harvard postdoctoral fellow Richard Russell and his team clinically identified super recognition in 2009. If there is a bottom 1–2 percent, they theorized, there is likely a top 1–2 percent. Those are the "supers"—the super recognizers. And while they aren't perfect at face recognition, whatever that means, they perform at the top end of the scale compared to the rest of the tested world. And the top 1–2 percent of the top 1–2 percent are going to look significantly different than even other super recognizers. As we'll see in chapters 3 and 4, supers are extremely good at a wide variety of face recognition skills: matching images to people or other images; aging people over time; and recalling people in different contexts and identifying them as the same. While super recognizers do not have a photographic memory for faces, and they actually do sometimes forget a face, they do it much less than everyone else. Super recognizers recall faces independent of the depth of the interaction or relationship they have had with someone. Most of us recognize the faces of those we know and love better than anyone else. We are more likely to recall those with whom we have shared a powerful experience. For supers, even a brief or fleeting or non-interaction with someone is often enough.

Recognizing faces feels like an act of memory, but it isn't, exactly. We process faces differently than almost any other thing that we see, though there is some overlap with cars and possibly buildings. First, we take in visual information via our retinas. It is passed to the occipital cortex, which interprets all incoming visual material and sorts it for further analysis. The occipital face area processes discrete features: eyes, nose, mouth. Then the fusiform face area configures these various parts to determine if it's a face, and if so, the superior temporal sulcus tries to make sense of emotion while also tracking lip movement to interpret speech.[6] All of these parts work together, and something can break down at any stage even as other parts are working well. Such a breakdown does not mean that someone is bad at remembering: these are two different processes. They work together. Poor face recognition can seem like poor memory, and vice versa. And a strong memory for detail and other aspects of self-presentation can help compensate for poor recognition. Someone who has a great memory may be terrible at faces. Someone who is excellent at faces may have a terrible memory. But the greatness in one arena or another may obscure the challenges in other arenas. Scientists have studied these regions through a variety of approaches, including MRIs that track what happens when human and non-human animals look at faces. As we'll see throughout this book, prevailing approaches to face recognition have changed and progressed rapidly in a relatively short amount of (recent) time. As we'll also see, the many pictures of people's brains provide only limited insight into face recognition and its lack. First, we had to identify face recognition as a category of variant ability. Then we had to figure out what that meant. Simply put: it's important to talk to people about what they do and do not recognize in the faces of others.

What Face Recognition Does

Face recognition is a wonderful and complicated neurological process about which we are learning more every day. It is also a deeply cultural, social, emotional, and human process. Faces are, as I've argued, a key part of how we make sense of others, build relationships, communicate, and make judgments. Recognizing faces helps with interpreting emotion. It tells us something about where people are looking and what they might be looking at. All this gives us cues about how to interact with others and our surroundings. Recognizing faces can help us recognize social cues about how to act and what to do in a given situation and with a given person. We spend quite some time on how we present our own faces, and we imagine we know all sorts of things about others based on their faces. The face is both a thing and a collection of things and feelings and ideas. The face has a history, and that history is changing. Face recognition, and the invention of the face recognition spectrum, is part of that history. The naming that emerged, with categories of "face blindness" and "super recognition," helped people understand something about themselves and how they make sense of others by giving a name to experience, and in so doing, creating new kinds of experiences. This is true of all categories, but there is special resonance to the face—because our faces are, or at least we think they are, who we are.

One of the themes that emerges in this book is that in the case of face recognition, naming helps. Categories help. We will explore the emergence of the particular entity of face recognition and the categories that lie along its extremes, and we'll discuss the challenges around naming and inventing something that is both deeply tacit and hard to identify. We will also look at the stakes for that process and what it meant for people to be able to narrate and understand their own lived experiences. As part of this examination, we will address bigger-picture questions regarding what it means to have a diagnosis, and whether a diagnosis in and of itself indicates patienthood. We will also question what it means to have a condition that you don't know is a condition until you do know, and what that knowing does to the condition, and to the one who has it. Along the way we will experience a great deal of relief, as well as aha moments, and, even still, a good amount of frustration and challenge. We'll consider adaptations and compensations and even as we try to think of the extremes of face blindness ("prosopagnosia") and super recognition together, we will be clear: the two are not the same. And one is much, much easier to hide.

A recent letter to *Slate*'s "Care and Feeding" column shows both the challenges of face blindness and the limits of naming. It's a column in the age of "AITA" ("Am I the asshole?") so it's also designed to highlight that people are terrible and will reliably behave terribly. In the letter, a severely face-blind woman revealed her son's ongoing hurt and anger that his mother never recognizes him. From a reader perspective, the son could display significantly more empathy and understanding. Yes, he's definitely the asshole here. At the same time, the letter—and the writer's anguish—highlights just how hard it might be not to be recognized by your own mother. And, equally, how hard it might be not to recognize your own son. Naming helps, but only so much:

Dear Care and Feeding,

My oldest son, "Michael" (26), recently sent an e-mail to both me and my husband that said he no longer wished to have any contact with me. His principal complaint is that I "don't respect him" because I "never remember his name." I have severe Prosopagnosia. I can't even reliably recognize my own face in the mirror, and have a lot of trouble recognizing people without intense study. So, while I can remember that Michael is the oldest of my three children, when I came down the stairs and looked in the living room, I often couldn't tell if the person sitting in the chair was Michael or one of his siblings. And yes, I would guess wrong on numerous occasions. And I always apologize when I get it wrong and am generally ashamed of how I can't manage something that seems so simple to everyone else. I want to fix this. I do love my son, and I want to tell him that I care for him and do respect him. But my last few attempts to do so before this e-mail just made things worse, and now he doesn't want to hear from me at all. How can I fix this?[7]

Reading this letter, it can be hard to understand what it's like not to recognize one's closest relations. That's the point: many of us, including "Michael," just can't understand. That doesn't mean we can't empathize. That doesn't mean we can't try to be generous. Michael drew all the wrong conclusions from something his mother could not change or do anything about. His mother knows that but still feels terrible. This is an extreme example: the mother is profoundly face blind, and her son's reaction is extremely harsh—one might say cruel. Most face-blind people encounter more understanding than this when they reveal their condition.

From a practical perspective, poor facial recognition means that it is very hard to identify particular people when you meet them; know which one is your own child if you are picking them up at the end of the day and they've changed clothing; go to the right table at a restaurant; make small talk when someone you know stops you to say hi; identify if a wave is for you or someone else; respond appropriately when someone comes in for a handshake or hug; not be on your guard all the time. *All the time*, as profoundly face-blind writer Heather Sellers has chronicled.[8] Likewise, as we'll discuss, face-blind psychiatrist and writer Oliver Sacks sometimes would see an image of himself and say hello.[9] That's an amusing anecdote that's a lot less funny when we think about the unremitting disorientation and confusion inherent in not being able to recognize oneself, let alone anyone else. Daily encounters can be fraught. Professional relationships can be strained. Some forms of entertainment are simply not entertaining because they are so hard to follow. Watching television is really difficult: when so many of the actresses look similar—by design—it's hard to follow the plot if you don't know who's who. Picture a group of the young, thin, white, light-haired women who dominate our screens: it can be hard even for non-face-blind people to determine who's who. But it's harder for face-blind people. If the characters are interchangeable, the plot is unfollowable, especially if it involves a double agent or double cross. A disability-rights case can be made for labeling characters onscreen; this particular form of captioning could help face-blind people enormously. Superhero stories, with their emphasis on bright costumes and the clearly delineated plot transitions, do a lot of the distinguishing work in ways other than facial, making them one of the preferred genres for those for whom face recognition is difficult or impossible. Another perhaps surprising exception to the challenges for the face blind in watching characters onscreen is the sort of show that features clones and identical twins: these shows do a lot of work at the level of dialogue, costuming and hair, gait, and even musical scoring to signal who's who. In a fascinating twist, face-blind people may actually be at an advantage in watching shows like *Orphan Black* and *Battlestar Galactica* because they are already experts at reading these distinguishing clues.[10] Adaptations work in interesting ways.

The face blind also generally have good compensatory mechanisms, though they often feel lost when trying to deploy them. Heather Sellers has chronicled her experiences with face blindness in depth, and she offers a deeply personal glimpse into her life without a diagnosis and her life with one. She describes herself as a highly social person who is good at creating instant intimacy. She greets everyone as though they are her best friend . . . because they could be.[11]

That's one approach. Another is to avoid greeting anyone in person at all, saving embarrassment, awkwardness, and a sense that everyone else has a superpower while you are deeply inferior. Sometimes face-blind people express that as well. For supers, it works differently. They too learn to fake it, but in their case, they fake *not* recognizing people. Their abilities can lead to awkwardness, as we'll see, and they too may either retreat or become hypersocial. Extremeness is itself a commonality, even if many other things about the extremes are different. Both extremes are also linked by the centrality of the face to the way they encounter others.

Facial recognition does not exist in a vacuum. We all recognize faces differently in different contexts and with different cues, even for those of us who can't do it at all. In some ways, actually, it's easier now to get by without recognizing faces, especially if you spend a lot of time interacting with others online. (As I write this in the shadow of a global pandemic, faces are often masked and sometimes unrecognizable in public, and most encounters with others are on digital platforms that provide names below the faces. As most of the world got Zoomed-out, a small group of people quietly celebrated the ability to always know to whom they were talking. For face-blind people, Zoom was an unlooked-for, unasked-for gift that gave the elusive possibility of recognition.) Recognizing people has never been harder, given the number of people in the world, and it has never been easier, given how much we encounter others on screen as labeled faces. But also, there are a lot more faces because there are a lot more people now. And we encounter them a lot more both in person and through media.[12]

We can know people don't recognize faces from their testimony. But there are a lot of ways to be bad at recognizing faces. So: Is face blindness a proscriptive term or a descriptive one? Is it a diagnosis or an explanation? The answer is yes: it is all these things. Face blindness, or prosopagnosia, is a specific term generated in a specific moment in history with specific reasons, narratives, and causes. It has to do with specific activity in a specific part of the brain. And maybe some of those people in the past had that activity in their brains, and maybe not. It's interesting, of course, to speculate as to whether some of these case studies with associated recognition challenges were actually examples of face blindness. Many people have made precisely those speculations. Others refuse to.

I, as ever, say yes, and.

What You Will See Herein

While people are the heart—and the face—of this story, it is not driven by a discrete cast of characters that we follow throughout the book. Instead, it converges around faces as historical object, biological entity, technological medium, mode of encounter, a presence, and a lack. Faces are a kind of thing, but they are also, I show, a process: the stakes for what they mean and how they are understood have been established and changed over time, and recognizing faces, or failing to recognize them, is itself a constructed and meaningful category that includes a diagnosis (eventually), a spectrum (evolving), and a lot of experiences (ongoing). There are overlapping moments and a lot of messiness, because history is messy and people are messy, and while we discipline these stories, that doesn't always work. I've chosen to keep the messiness and the overlapping time frames and disparate characters, following the face and its recognition around rather than the people who studied it or live with its consequences. We will meet these people, and they'll keep popping up, but this isn't a biography of a person or a place. It is, in its way, the biography of a concept: face recognition. Which means that this is the story of a diagnosis, and the processes by which we got there. While this is not an ethnography on living with the extremes of the face recognition spectrum, I ground these claims in personal stories and lived experiences through both recorded history and personal interviews. With them, I think about what kind of entity face recognition extremes are: Are they disabilities? Super skills? What is the opposite of a disability, and how do we make space for the interplay between adaptation and lack? This book marks changes over time; it also unveils hidden systems, showing workings that become visible only in their failure. I track the science and its infrastructures: publications, funding, positions, strategies. But I also make a plea for the humanities, arguing that face recognition becomes visible and meaningful through experience, cultural production and processes, and, simply, talking to people.

The pull toward a search for origins is powerful, and a longer view can help frame current understandings of face recognition. I offer here two beginnings: the detective story that I did not want to write, a deeper history of what might have once been face blindness through a close look at the nineteenth- and twentieth-century mind sciences literature. As I explore in chapter 1, before prosopagnosia was, in the language I use throughout the book, *invented* in the twenty-first century, some people had trouble recognizing faces, so much so that others noticed. And made note of it. That doesn't mean that it was face blindness exactly, but it was something notable. Which means, at the very least, that recognizing faces has always (or at least for a long time) mattered. And *not* being able to recognize faces, regardless of exactly why or how, has long had stakes. The whys and hows of those failures—clinically, medically, culturally—are not the right questions. We can't ever know what was "really" happening in x's brain or with y's encounter with faces. But we know that faces and the ability to recognize them mattered then, just as they matter now. And that not recognizing them was a challenge then, just as it is a challenge now. Though the kind of challenge it presents has, perhaps, changed: in scope, in scale, and, as ever, in context and stakes.

Chapter 2 begins not at the chronological beginning, but when face blindness, as a modern experiential category, was invented—not with its clinical naming in 1947 but in the first two decades of the twenty-first century. Here I explore the creation of face blindness in the modern era, tracking its clinical framing in conjunction with its mediation to produce the discrete category of prosopagnosia. I think about the role that easy access to self-diagnostic measures, including online quizzes, plays in its development and dissemination, alongside the high-profile discussion of living with prosopagnosia from Oliver Sacks and Chuck Close. While I draw on these men as players in the game, I am reflexive about the stakes for "Great Men History" and think carefully about how I engage with that narrative even as I complicate it.

Chapter 3 builds on the expansion of knowledge around face recognition as a category rather than tacit and unnamable knowledge to consider narratives about super recognition. In this chapter I tell the story of how super recognition was invented and the stakes for reframing face recognition as a spectrum. I offer narratives based on individual experiences, drawing on my interviews with face recognition researchers Richard Russell and Josh Davis, as well as the first clinically identified super recognizer, Jennifer Jarett. I think about the overlaps and points of departure as they all describe similar, but meaningfully different, stories. I also consider my own role as chronicler and historian in taking these individual stories and establishing an official history.

Chapter 4 turns to the lived experience of inhabiting the category of super recognition. I explore attempts to leverage and monetize super recognition, thinking about the different stakes for naming something called "blindness" as opposed to something called "super." I also stop to ask about the losses in the "super" framing and argue for thinking of the two extremes together rather than always in opposition. Chapter 5 offers a Cold War link in the biometric surveillance chain, exploring analogue attempts to make people into better recognizers well before the emergence of super recognition. I turn to the archive of a little-known face recognition researcher, Peter Bochmann, the head border guard at Checkpoint Charlie in Berlin. Bochmann was committed to eliminating border fraud, and to that end, he developed a number of ways to test and train face recognition. I show the intertwined histories of surveillance and face recognition, leading to chapter 6, which shifts to digital attempts to build and leverage face surveillance on a global scale.

Chapter 6 explores face recognition technology and its relationship to machine learning as Cold War technologies designed to mimic the actions of the human brain and improve upon them in the project of tracking people, their words, and their actions. I chart how people and their biases are always integral to machine heuristics. Face recognition software was initially an attempt to learn about how people recognize faces by observing how machines do it. But machines do algorithms. People reason. Product, which in this case is recognition, is not the same as process. And there are deep costs to imagining that those machines work under the same processes as people. Even as we improve its algorithms and limit its biases, the software remains embedded in larger structural systems of surveillance that, in the end, sort people. And it does so based on the biases and goals of those developing and implementing the systems. And systems can be very hard to see. As I explore in chapter 7, systems work to obscure their own workings, making it difficult to identify what they do until they fail. I look at the ways that face recognition is itself one such system, and in this highly theoretical series of thought experiments, I argue that it is the deeply tacit nature of this process that makes it so hard to identify in success and in failure.

In a coda to the book I present an oft-repeated story from antiquity about fevers that made people unable to recognize those they know. I won't call it face blindness, but I offer it as a third, much older, possible origin story, both because it is powerful and also to highlight that when it comes to these sorts of beginnings, we can never quite know where something starts. But we can examine what it means.

The idea of face blindness can be terrifying, the stuff of horror films and dystopias, a world in which you can't recognize anyone you once knew. A world in which you cannot be recognized. A world in which the monster suddenly has a blank face; a world in which anyone can be the monster. The face, in this version of face blindness, is detached from personal experience. It becomes a kind of unrooted thing, seemingly unrelated to relationship itself, an unmoored, unidentifiable elusive key to identity. But that detachment is true on both ends of the spectrum. Let us invert the problem, the question, the terror: What would it look like to live in a world where knowing people is unrelated to recognizing them? This, in a way, is the world of both the face blind and the supers; we can see how the rhetoric around these conditions draws from the rich tradition of theorizing the meaning of the face. Such theories provide important insights into the histories and social meanings of face recognition; they shape the rhetoric around face blindness and super recognition and what they mean. Though the research has progressed a great deal, the imagination is still limited. Yes, we have the spectrum, but we still understand super recognition and prosopagnosia as binaries in sharp opposition to each other, and we still don't examine those extremes together to learn about the ways that worlds can be similar as well as different.

1

Thinking in Cases

A Somewhat Failed Search for Origins

The problem with studying historical cases of a bodily phenomenon is that one has to be careful. But still, it seems useful to know that this long-unnamed, uncategorized, and undefined phenomenon, this most tacit of bodies of knowledge and most fundamental of relational responses, has historical traces. And yes: there are significant and historically robust limitations to retrospective diagnosis. Given how deeply bodily experience, medical infrastructure, and diagnostic practice are intertwined with broader cultural and structural contexts, applying current ideas of what a condition might be to the past simply doesn't work. It's not just that we can't know exactly what was going on given the different ways that people were evaluated, the different framing of what evaluation even meant, and the significance of the often limited records and answers, though all of these factors are highly relevant. But meanings change over time. Contexts change over time. Processes that once worked in one way work quite differently now. What does it mean to be unable to recognize faces at a time when there were far fewer faces in most people's lives?[1] And yet, the temptation to revisit historical cases is irresistible. The pull to origins is strong.

But we can be playful: here I offer at least two places to start, and I'm sure there are more. This is one version of an origin, in the nineteenth century. But it is limited, because there is also a different kind of origin, one I call invention, that emerged when face recognition and its lack became an identifiable category of experience—and that happened much later indeed. The origin is situated here in two historical cases described by practitioners; the invention came from the descriptions of those with lived experience. In these historical cases, we hear from the patients only through their doctors and only as individual cases. But these two cases can help us understand something about face recognition and its lack, even if understood and experienced in a different time, place, and way.

I am ambivalent about retrospective diagnoses. I recognize the potential to make false claims and lose both context and meaning. But there are losses if we reject them entirely; it is possible, in fact, that there are things we can know about the past and the body.[2] There is value in trying to understand what was once known, if perhaps in a different way. Doing so creates space for what one day may be a fact about bodies that cannot yet be understood. The very question insists that diagnosis is a limited way to think about bodily knowledge and bodily capabilities. While diagnosis often accompanies disease and disability, identifying limitations, impairments, and the need for accommodation, it can also, in my framework, highlight how such categories can be mobilized in useful and advantageous ways. But only when we can know them—only when we can see them. Diagnosis itself can be valuable with or without disease. It can serve to create visibility, marshaling both resources and relief for those who can now name and instantiate experience. It also can take on a great deal of power and a life of its own. We must be attentive to context, but what do we lose when we declare certain concepts or conditions to be unknowable, even when there are concepts or conditions that we know?

We must, when thinking in and across cases, ask: What can one case from antiquity (see the coda) tell us? What can one case from antiquity tell us alongside four cases from the nineteenth century? And four cases from the twentieth? And what can these cases tell us when we unearth, produce, discover, or describe hundreds—or thousands—more cases in the twenty-first? To think in cases, in John Forrester's terms, or rather to reason in cases, we extend the similarities across these examples to define the particular and the exemplar as a mode of epistemology, a way of thinking.[3] One—and indeed one and one and one—is enough to reason in cases.

But first the examples have to be named. Through naming, recording, contextualizing, and comparing, they have to become cases. That, too, as medical sociologists have shown us and as I discuss in chapters 4 and 6, is very much a process.[4] And that process in turn mobilizes other processes and communities and affinities, and—deliberately—advocacies and expenditures. It doesn't always work out that way: some individual cases get dismissed or marginalized rather than generalized and exemplified.[5] For Forrester, such failures might happen because of a category error, in which "case reasoning" is misrecognized as the statistical or taxonomic reasoning so beloved in the nineteenth century and so central to the workings of medicine through the twenty-first.[6]

Can we reason from cases with faces? Is that the answer to not engaging in egregious retroactive diagnoses? Only sort of, because these are only *sort of* cases—"cases(ish)" as I call them here until we get to what looks like a modern case. Indeed "the case" itself is a historical construct, requiring an understanding of "the individual" and "the self" as connected to medical and psychological experience. Without the self, there are examples and exemplars, but not cases.[7] And language itself is tricky: the term "face blindness," as we will see, was first used in 1899; prior to that, we need to work from descriptions and examples, and we draw some connections and maybe conclusions.[8] Where we cannot reason, we can describe, and where we cannot describe, we can recount. Then, once we have described and recounted, perhaps we can think across time, across experience and context, and in so doing, just a little bit, we can reason.

Case(ish) 1: Arthur Ladbroke Wigan, 1844

British general practitioner Arthur Ladbroke Wigan wrote a treatise on insanity and duality of mind in 1844.[9] His theory of two independent hemispheres that mostly collaborate but may occasionally clash held (and continues to hold) wide purchase and found its way into cultural productions like Robert Louis Stevenson's *The Strange Case of Dr. Jekyll and Mr. Hyde* (1886).[10] Among Wigan's examples was a case of a "gentleman of middle age, or a little past that period" who "lamented to me his utter inability to remember faces." Specifically, "he would converse with a person for an hour, but after an interval of a day could not recognise him again." This was true not just for strangers and acquaintances: "Even friends, with whom he had been engaged in business transactions, he was unconscious of ever having seen." He developed other strategies of identification, waiting until "he heard the voice," and then "he could recognise men with whom he had constant intercourse." Wigan probed, finding that after "I inquire more fully into the matter, I found that there was no defect in vision, except that his eyes were weak, and that any long-continued employment of them gave him pain." The gentleman "was quite determined to conceal it, if possible." He remained convinced that the problem depended "solely on the eyes" rather than anything related to his brain. The patient was resistant to sharing his challenges and indeed showed "a strong desire to conceal his defect from the world."[11]

This is, in fact, a rich and resonant case, for both the clinical approach to, and the lived experience of, difficulty in recognizing faces. Much of it may sound familiar, and perhaps it should; perhaps this is a lot like face blindness. We have a man who can identify the person immediately in front of him, but once that person has gone, no matter how long the two interacted, the man cannot subsequently recognize him. This is true of strangers and friends, including those with whom he had regular contact. He developed strategies to compensate for his deficit, relying on voice cues to serve as a means of identification for those he knew. The problem lay not, according to the testimony of his physician, in his eyes, which exhibited only standard weakness. And, importantly, the patient did not accept that his eyes were largely functioning as they ought: in the absence of any better or more commonly recognized explanation, a deficit in sight made the most sense for this enduringly hard-to-identify set of experiences.

We know from this case that the gentleman experienced difficulty remembering faces, and we know that he considered that difficulty a challenge. This is worth noting: so many of those with face recognition deficits have repeatedly expressed how hard it was to understand that these deficits were real, and to identify that these deficits were worse, in a meaningful way, than the general trouble most people casually experience in remembering faces. Wigan's patient is remarkable in the clarity he had that this trouble exists; that he had developed compensatory mechanisms; what, precisely, those mechanisms were; and how and why they worked better in the case of those he knew well (and those whose voices he could remember). He also continued to attribute his challenges to his eyes rather than his brain, relying on his own bodily experience and knowledge rather than his doctor's perspective on the roots of the problem. He did not want others to know about his problem, and, as far as we know, he succeeded in hiding it. Wigan did not explicitly overrule his patient or offer alternative explanations; he merely noted that the weakness of the patient's eyes did not seem to account for this particular recognition difficulty. Both doctor and patient told us what they knew and how they knew; they did not tell us what they did not know.

This case helps us understand a great deal about this patient in particular, and his own experiences, strategies, and convictions. We get a sense, but not much more, of what it was like for him to live with this difficulty in recognizing faces and how he felt about it. We learn about his own relationship to this difficulty. His desire to hide his condition may have spoken to a sense of embarrassment or shame; as we have seen, others with similar difficulties later expressed similar sentiments. We might also know something, in conjunction with other cases, about this case more broadly. But life in 1844 was different than life in, say, 1996 or, once the face-blindness floodgates opened after that, life in the new millennium.[12] So I want to be careful here but also at the same time acknowledge that we can think across cases, even knowing that the cases might be different—must be different, because of their context—to consider what might be the same. That approach is valuable. It is equally valuable to consider where these cases and descriptions diverge.

Case(ish) 2: Antonio Quaglino, 1867

In an 1867 paper Italian ophthalmologist Antonio Quaglino (1818–1894) offers a much more detailed case than Arthur Wigan of a patient with face recognition deficits.[13] Psychologists and cognitive neurologists Sergio Della Sala and Andrew Young comfortably identified the patient in question as having "a clear prosopagnosia coupled with an inability to recognise buildings in his own town and colour blindness."[14] Quaglino's patient, a 54-year-old bank clerk from Turin known as Mr. LL, experienced these deficits following a stroke in the right hemisphere of his brain; like the example we will cite in the coda, this patient acquired the deficits after a traumatic medical event. Prior to this stroke, he did not exhibit any notable difficulties in face recognition, building recognition, or color differentiation. While I might be more careful than Della Sala and Young in applying the label prosopagnosia with such confidence, I am struck by their assessment in their translation of Quaglino's paper on this topic that "Quaglino's case is of special interest because it anticipates and bears so directly on points that have arisen in later discussions of the topic of acquired impairments of face recognition."[15] This too, is a way of reasoning in and across cases.

Quaglino and Borelli's patient, Mr. LL, fell to the floor unconscious on February 28, 1865. He recovered after several days but continued to have weakness on his left side and blindness in both eyes. These symptoms improved over time; when Mr. LL came to Quaglino, his near and far vision was excellent, as was his bodily movement. Although his heartbeat was arhythmic, he seemed at first to have made an excellent recovery from his stroke. At first. He had come to Quaglino for a less visible vision deficit: he had lost all ability to distinguish color, buildings, and faces. Quaglino recounted that "what dismayed him most was that no sooner had he got out of bed, he saw the faces of all people as doughy and whitish." This lack of definition was compounded by problems with color perception, as "of all colours he could only distinguish white and black." Quaglino confirmed this finding with a test, noting that after "having presented him with a series of large characters in various colours, in fact, he failed to name any of the colours." A lifelong affliction was quickly ruled out: upon being "questioned as to whether this defect was congenital, he confirmed that before his stroke he had always had a clear and distinct perception of all the colours." The doughy and whitish faces were not just dismaying in and of themselves; they proved particularly hard to identify and parse. Mr. LL "also noticed that he had completely lost the ability to remember faces, the facades of houses, and familiar scenes." The problem was quite general: "In a word, he had lost all the shapes and the configurations of things." Quaglino ends this section with a poignant quote from Mr. LL himself, writing that "I remember . . . the names of my friends, I recognise them when I meet them, but as soon as they have turned their backs to me, I no longer recall their features."[16]

Quaglino continued his report, specifying that he found no major damage to the eyes themselves. He concluded by theorizing the presence of "damage to the visual centres on both sides." He was careful to note that "it is impossible for us to determine whether the damage that gave rise to the deficits we described took place in some visual centres (e.g. superior colliculi), or rather within the convolutions of the cerebral lobes that attend to the secret functions of cognition." Given "the absence of clear data we would rather not venture further speculations." Building on Franz Joseph Gall's theories of brain localization, he concludes with the contention that this case "add[s] further support to the notion of the brain as a collection of specialised centres. Following lesions to these specific areas of the brain, alterations, weakening or interruption of specific functions can occur, while other functions remain intact."[17]

It is very helpful to know the limits of what Quaglino could know from this case, which should guide our own analysis of what it reveals. From the report we know that Mr. LL was no longer able to recognize buildings, colors, and most upsettingly, any faces at all, including those of friends. In this case, we have the patient's lived experience: dismay. We have a probably causal origin: the stroke. We can rule out at least one factor: damage to the eyes. We can surmise another factor: damage to some part of the brain. We can say from the evidence presented that the problem did not get better. And we can ask: What did Mr. LL mean when he said he recognized his friends when he met them?

Perhaps, like the gentleman in Wigan's study, Mr. LL was relying on other cues, like their voices. It was not their faces he recognized, since they, like all faces, appeared "doughy and whitish" to him, and indeed, "as soon as they have turned their backs to me, I no longer recall their features."[18] This is familiar territory for the face blind, and Mr. LL's specificity is notable and descriptive. While people were in front of him, he could indeed recognize and make sense of their features. And once they turned away, their features were gone. Many face-blind people report that they do in fact see both individual facial features and the face as a whole when they are looking directly at them. They just can't seem to remember them.

Case(ish) 3: John Hughlings Jackson, 1872

Quaglino's beautifully laid-out example was on its own for another five years, until British neurologist John Hughlings Jackson (1835–1911) described a patient in 1872 who "at one time . . . did not know his own wife."[19] He lost his ability to recognize places and people, and he got easily lost even in familiar surroundings like his own neighborhood. He had damage to the left side of his brain, which Hughlings Jackson attributed to a lesion in the posterior areas of the right cerebral hemisphere. Hughlings Jackson's exposure to these symptoms was useful four years later when he was presented with another patient who "often did not know objects, persons, and places," a problem that gradually grew more and more pronounced. Based on his prior experience, he accurately diagnosed a right posterior tumor, which was later confirmed at the patient's autopsy.[20]

Hughlings Jackson learned from one case to apply it to another. Both patients presented with similar and unusual symptoms, allowing him to both make a diagnosis and reason more broadly around the neuroscience of vision and its location in the brain. He theorized that posterior lobes "are the seat of visual ideation" with "the right posterior lobe" as the 'leading' side" and "the left the more automatic."[21] Hughlings Jackson framed subsequent discussion on cerebral asymmetry, though he could not specify different visual disorders. Importantly, he clearly articulated the difference between sight deficits and recognition deficits, work that later discussions would build upon. As we have seen, both Wigan and Quaglino noted that the deficits their patients demonstrated were not linked to vision per se, but Hughlings Jackson incorporated these distinctions into a broader theory of neurovisuality.[22]

Case(ish) 4: Jean-Martin Charcot, 1883

French neurologist Jean-Martin Charcot, perhaps best known for his work on hysteria, presented an even more explicit case of face blindness in 1883.[23] However, his patient, M. X., did not suffer with deeply impaired color recognition and did not seem to have the left visual field defect that had been noted in prior as well as subsequent cases. In other ways, M. X.'s experience resonates deeply with the limited collection of examples to date.[24] His sudden loss of visual imagery did not seem to be precipitated by a stroke or other injury to the brain. His experience was neither congenital nor tied to a related medical event as far as he or Charcot knew; his "mental imagery [*vision intérieure*] has disappeared completely." This was a particularly dramatic shift, given "that this mental imagery that I lack today used to be possessed by me in no ordinary degree" and in fact "is still possessed by my brother, professor of law in the University of X, by my father, an oriental specialist known in the scientific world, and by my sister, a very talented painter." He wrote that "I used to have a great ability to picture to myself [*représenter intérieurement*] persons who interested me, colours and objects of every kind, in a word everything that is presented to the eye." He "used this ability in my studies: I read what I wanted to learn and when I closed my eyes I could clearly see again the letters in great detail"; and he applied this ability to other arenas in his life. "It was," he wrote in a letter to Charcot, "the same for the appearance of people, or countries and cities I have visited in my long journeys, and, as I just mentioned, for every object I had seen with my eyes." This all disappeared, and quite suddenly: "Today, no matter how much I want to, I cannot image [*représenter intérieurement*] the features of my children, of my wife, or any everyday object."[25]

According to Charcot's presentation of the case, M. X. understandably panicked about this dramatic loss, and indeed "was led to doubt his sanity," though he quickly became "reassured by noting that he could still see shapes and colours even though he could not remember them." Even more comforting, he quickly adapted, drawing on other skills as "he realised that he could continue to direct his business affairs by relying on other forms of memory." The problem, as we have repeatedly seen, was not with his eyes: "Ocular examination by Dr. Parinaud had revealed nothing which could account for M. X.'s problems, though he was noted to be myopic and to have suffered a slight loss of sensitivity to colour." Not only did he lose color but anything in his visual memory, including texts that he had once "been able to 'read' . . . from an image of the page." To compensate, "M. X. found himself having to rely on auditory memory and 'inner speech,' which was for him a novel experience."[26] All of M. X's visual experiences disappeared; he wrote that "with this loss of visual mental imagery [représentation intérieure], my dreams are correspondingly changed." Whereas "in the past I always had visual perception in my dreams," he lamented that "now I only dream in words."[27]

M. X.'s loss had a noticeable effect on not just his daily life but also the way he experienced the world. He wrote that "no longer being able to form visual images [me répresenter ce qui est visible], and having completely kept my abstract memory, I experience daily astonishment at seeing things I must have known for a long time." This shift profoundly impacted who M. X. felt he was and how he acted: "With my sensations, or rather my impressions, being indefinitely novel, it seems to me that a complete change has come over my existence and of course my personality [caractère] has changed notably." He lamented that "I used to be impressionable, enthusiastic, and I had a vivid imagination; today I am calm, cold, and my imagination can no longer lead me astray [ma fantaisie ne peut plus m'égarer]." Indeed, "a remarkable consequence of the loss of this mental ability is, as I have said, the change in my personality and my impressions." More specifically, "I am much less affected by sorrow or grief. I can tell you that after losing one of my relatives, with whom I was very close, I experienced a much less intense grief than if I had still been able to image the features of that relative, the stages of the illness he went through, or especially if I had been able to picture to myself [voir intérieurement] the outward effect of this untimely death on the members of my family." With this loss, he was a different person.

Like others, M. X. failed to recognize familiar landmarks and found himself getting lost in once-known environments. This failure included not just buildings and spaces but people, including his intimates. M. X. ceased to recall how his wife and children appeared; even when they were in front of him they seemed unfamiliar. As indeed, did he: he recounted standing in someone's way in public and offering apologies, only to realize he was look-ing in a mirror.[28] Of course, as we heard from M. X. himself, "the impact on M. X.'s life was dramatic" and quite challenging.

Let me pause here and reflect on how poignant these experiences are. They were powerful coming from M. X., and they were powerful coming from Oliver Sacks, who wrote about reminding himself that the person with the large ears waving at him was in fact . . . himself.[29] It is worth remembering that sometimes we can draw conclusions across time, space, and culture. M. X. presented with some sort of facial recognition difficulty in 1883; Oliver Sacks wrote about that same problem more than one hundred years later. But they said almost exactly the same thing. What does reasoning in cases tell us about this? At the very least, that they both had trouble recognizing themselves and others, and that it seemed to be a problem with recognition more broadly. And that it was really hard.

For Charcot, M. X.'s case supported his assertion that visual recognition and memory is distinct from other forms of memory. As Della Salla and Young pointed out, Quaglino was more specific in recognizing this form of visual impairment as distinct in its own right rather than subsuming it under a larger category like visual ideation loss. They connected his theoretical ap-proach to his commitment to brain localization as suggestion by German neurologist Franz Joseph Gall; although Gall's phrenology was by this point less respected in scientific circles, his underlying mechanism, Quaglino rec-ognized, held great value and explanatory power.[30] While Quaglino's approach has deep consonance with today's theories of face recognition deficits, his work did not factor into most studies and approaches and remained some-what marginal to the field until recently.[31]

I've been careful about terminology, avoiding the terms "face blindness" and "prosopagnosia" to categorize these historical cases. The patients and doctors and record-keepers did not use this language, and nor will I. But "face blind" does emerge in 1899, likely the first and, until 1947, only use of "face blind" in the literature.[32] This raises the question: Now that face blindness has become an actor's category (by which I mean something that the histori-cal figures that I study used), am I free to use it in my descriptions and discus-sions? Most historical approaches would say yes, though they would caution against drawing a straight line between this face blindness and later ones. The language may be consistent, but the experience and indeed the phe-nomenon itself may well be entirely different.[33] So the face blindness of the

1899 "Transactions of the Michigan State Medical Society" may or may not be the same as the face blindness of 1947, which may or may not be the same as those that follow. The later cases, however, were in many ways co-constituted and deeply intertwined, making it easier to think of them together and analyze their relationships. We must be more cautious around the term at the turn of the (twentieth) century, and we must be attentive to its particular language and structure.

For instance, a number of symptoms were considered forms of blindness then. In this moment of media innovation, sight and its lack framed much of the way that deficits and advantages were labeled and understood. As we discuss in chapter 7, there was a cluster of blindnesses: color, word, and, now, face. The pervasiveness of these language choices is not accidental: sight is of course itself both a medium and a system that engages with other media and other systems, all of which were under enormous pressure to function in an ideal state to produce productive capitalist citizens who could see and who could work. Sight itself became not just a skill but a metaphor for understanding, enabling, and executing. Deficits in these arenas were not just relative weaknesses but absolute lacks, or blindnesses. It was an either/or designation, with no room in the framing for variation or compensation. You could see and do, or . . . you could not.

Case(ish)5: David Ingles, 1899

The 1899 "Transactions of the Michigan State Medical Society" report made these blindness links explicit. In his brief report, Dr. David Ingles of Detroit presented the case of what he called "moral imbecility" in the context of a murder trial. In building his argument as to the fundamental untrainability of certain deficits, including in some cases morality, Ingles drew analogies between color blindness, word blindness, and speech aphasia. He offered that a "color-blind man is no rarity." While he "is in all other respects normal," "he is utterly unable to distinguish some, or in rare instances, all colors." This is not a function of "an undeveloped capability," but rather "the absence of a capability common to normal men." It cannot be learned: "No amount of training of effort, of attention, will enable the color blind man to distinguish colors." As such, "no one . . . rightly holds a color blind man responsible for not seeing colors." For Ingles, "this illustration" is but "the most common" among many an "instance of limited defect in a mental organization otherwise sound."[34]

It's a good and compelling argument. Ingles then personalized it, noting that he was "subject to an analogous difficulty." Specifically, he revealed, "I am face blind." He went on to state that "I can by no means recall to my mind the memory of a face." This was true for acquaintances and loved ones, as "I have no mental picture of the faces of those nearest and dearest to me." Similar to M. X., Ingles's visual deficit was almost total. He wrote that "I do not see faces in my dreams." And, like the other cases we have encountered, it was hard for Ingles, particularly socially and professionally, as "necessarily such a defect is a great drawback, for I am continually in difficulties from failure to recognize acquaintances." He tried, but "no training, no voluntary effort has apparently any power to correct the defect." He emphasized in emphatic terms that "it is again not a faculty untrained but a faculty lacking."

For those still doubting, he provided yet another example that "such a limited defect is possible" which "is shown by the phenomena of word blindness." In that case, "the memory for words as seen by the eye is lost." Once again, it is not the eye itself that is faulty, as "the memory of all other visual objects [is] retained." He then brought up "certain aphasics" who "have lost the ability to recall nouns while able to recall and use all other parts of speech." And, he concluded, "the list could be extended." All of "these and similar phenomena demonstrate the fact that the mental organization may be perfect in all respects," or close to perfect "save some limited and isolated function." "But," he emphasized, "as far as that function is concerned, no possible training can correct a defect which is as real and definite as an amputated leg."[35]

I found myself deeply moved by Ingles's framing—not just by the personal testimony from over a century ago that is so deeply resonant with so many others who came later, though that is itself poignant and almost haunting. But because it is *such a clear explanation and powerful argument*. I, like so many other people who are not face blind, struggle to understand the experiential mechanics. Not how it feels: I kind of get that, but rather, what faces actually look like to face-blind people. As an academic who studies faces, I have read all the explanations and metaphors, from the piles of Legos to the cheetah faces to line drawings that we reviewed in the introduction. They help, as much as anything can help someone understand something they cannot themselves experience, see, or inhabit. But Ingles's narrative is powerful in a different way. It neatly and cogently places neurodiversity in an easily accessible and value-neutral category. It's a century old. We can quibble with some of the language through twenty-first-century eyes. But as a way to approach neurodiversity alongside other ways of being in the world, it is excellent. It is a robust call to show why the categories of blame and self-improvement are simply meaningless; we can extend Ingles's logic to argue that people with what he calls deficits should in no way feel compelled to change themselves (because they cannot). Rather, it is the world that should and must change.

And in this, of course, he was right.

Ingles's paper is widely credited as the first use of the term "face blind." I have not found any earlier examples, but I admit I'm a bit skeptical: he used the term with such confidence and without any claims to originality. That may have been a strategic rhetorical decision on his part—he was a master— but it does seem like he was offering something that was already in wide use. Given the paucity of reported cases of face blindness, it's surprising that there would be any kind of use, let alone a wide one, but equally, Ingles himself only offered his case anecdotally in service of something else. That has its own frustrations, despite the richness of his description. Is this in fact not just the first use of the term "face blind" but also the first recorded example of congenital face blindness? It seems like there was no injury or discrete cause, though we cannot know for sure. He was not being followed or (pointlessly) treated for his condition, and without the particular imperative offered by the argument he made in this paper, we might never have known about his face blindness. That too, tracks: until quite recently, we could say this about most face-blind people: we, and they, would never have known. Never, in fact, did know.

And yet so much about the Ingles case seems so very stable. The inability to recognize even close intimates. The facelessness of dreams. The social and professional challenges. The lack of eye trouble. The fundamental inability to change the condition.

The name.

But it was nearly fifty years later that we got another example. We can reason that there were many other people with similar deficits and challenges. We can equally reason, because we know, that these people did not even realize that what they had might be categorizable or diagnosable. Actually, that's not quite right: it wasn't that people didn't realize it, it was that it *was not* in fact categorized or diagnosed. It was yet to be categorized, and so as a medical or neurological phenomenon, it did not quite exist. The cases to date were in the context of other events—sudden changes, major illnesses, dramatic injuries, theoretical arguments. Face blindness was, largely, a symptom, not a condition or a syndrome in and of itself. We can reason that though face-blind people, and face blindness, existed, until it became a category rather than a symptom, it was largely invisible. And that, for all these reasons, took a while. But luckily, when the next case was recorded, it was done in such depth and detail, providing a richness of experience and narrative that helped not just shape the research and conversation, but ensure that that research and conversation would occur.

Face blindness, or difficulty recognizing faces, has long existed as a lived experience. Through the nineteenth century, it existed as a symptom. As of 1899, it existed as a name. But it was not until 1947 that it came to exist as a diagnostic category. There were many other agnosias described along the way, including Hermann Wilbrand's 1887 case of a woman who lost the ability to dream following bilateral cerebral artery thrombosis.[36] Otto Potzl refined a series of case studies to describe Charcot-Wilbrand Syndrome (CWS) in 1928 as "mind blindness with disturbance of optic imagination," a condition of dream loss and the inability to recall images following brain damage. Mind blindness is like—but also not like—acquired face blindness, causing a loss of visual memory and recognition after a traumatic brain event.[37]

As many scholars have illustrated and as we discuss in chapter 4, naming and establishing a condition as a diagnostic category does a lot of work in both patient affiliation and agenda setting.[38] While the experience and indeed the symptom have a long and embedded history, the condition itself can only be said to emerge in conjunction with an institutional framework, individual experience, medical/expert attention, research projects, and communal advocacy and exchange. These did not all happen at once: creating a diagnosis, a disability, a disease, is, of course, a process.

While experience and symptoms and names can be studied, it is much easier to do that when there is a category. And a category usually takes more than a case. Categories both require case reasoning and case collection. An n of 1 is a story. An n of many can (but may not always) be an agenda.

Case 1: Joachim Bodamer, 1947

German neurologist Joachim Bodamer's (1910–1985) 1947 paper did just that.[39] While, as we have seen, it was far from the first clinical description of face recognition deficits, it was the most comprehensive to date, systematically chronicling two patients with similar symptoms and contrasting them with a third, using these examples alongside historical cases to offer a theoretical framework. Bodamer's report helped shift face recognition from being a symptom to being a syndrome in its own right. He also came up with a great—and enduring—name: prosopagnosia. *Prósōpon* is Greek for face/mask/person, and *agnosia* comes from the Greek *agnōstos*, meaning the state of not knowing or "unknowable." Prosopagnosia was framed to be a "new sub-category of agnosic disorder," that specifically would "delimit it from other partial forms of agnosia," namely the clinically well-established failure to recognize objects, faces, voices, or place.[40] In other words: face blindness.

The name caught on, and "quickly gained international approval" according to Ellis and Florence, who, in 1990, offered the first comprehensive English translation of Bodamer's German 1947 paper.[41] Bodamer offered two systematic descriptions of patients who presented with what he called prosopagnosia, building on prior research in agnosia and offering prosopagnosia as a specific form that could establish individual types of agnosias as "new, well-circumscribed, well-definable single forms" while using visual agnosia to learn more about the structures and hierarchies of operation within the brain itself.[42] His detailed definition of prosopagnosia drew on the category of agnosia generally while insisting on specificity for each type. He wrote that prosopagnosia is "the selection disruption of the perception of faces," emphasizing that this includes "one's own face as well as those of others. Again, the problem is not with the eyes, as faces "are seen but not recognized as faces belonging to a particular owner." He acknowledged the degrees to which someone might be affected, given that prosopagnosia "appears in varying strengths." It often came with other deficits but ought always to be recognized as its own neurologically unique syndrome; although it "appears in varying strengths and together with the most different forms of agnosia," it "can be separated from these from the outset." He defined prosopagnosia as "a well-characterized, strictly defined disorder, situated on one area of the sense, affecting on a certain part of the gnostic activity." In particular, "that of the recognition of faces."[43]

Bodamer was pushing hard for the specificity of prosopagnosia. He cited the work of those who looked at cases of this phenomenon in the past; most of those cases, he argued, involved object agnosia "in the foreground," causing the "disturbances of faces" to be "subsumed by object agnosia." The disturbances of faces were "noticed . . . only in fresh cases and along with strongly marked object-agnosic disorders." Over time, "prosopagnosic disorders disappeared along with other phenomena." Bodamer was less interested in "fresh" and possibly temporary examples, though he lauded "the observations of Charcot, Wilbrand, Heidenhain, and Jossmann" as being "especially valuable for the way they determine the extent of disrupted face-recognition in a particularly pure form."[44] Hoff and Potzl, he continued, did suggest face recognition deficits as a distinct agnosia, but they used the language of a disorder of memory for faces. Just as earlier researchers dismissed the issue as not a deficit of vision, Bodamer thought it was not a deficit of memory.[45]

Bodamer's first case was patient S, whom Bodamer observed when S was admitted to a brain injury hospital in 1844 for a serious bilateral brain injury.[46] S was blind for several weeks following his injury; while sight gradually returned, "all impressing of colour was missing—he could distinguish only light and dark, black and white." Poignantly, "he saw everything faded, strangely colourless." But the memory was not gone, as "he could imagine colours, and in dreams objects and events appeared in their natural colours." This was hard, as "he was deeply disappointed when he awoke to a colourless world." When Bodamer observed him, S was "completely colour-blind." Bodamer used this term to denote the complete lack of ability to distinguish color rather than the more common usage in red-green color blindness.[47] We can say that Bodamer's use—and S's manifestation—is a more accurate application of the term: most "color-blind" people, are, in fact, not really color blind.[48]

In addition to the inability to distinguish color, S also exhibited minor—and improving—object agnosia. He was able to recognize some but not all objects, failing in particular with those he had not encountered since he was injured. He drew on touch and deductive reasoning rather than visual analysis to determine what various items were, and he was consistently able to visualize objects and images in his mind, demonstrating an undisturbed visual memory.[49] His other weaknesses included reading, making sense of pictures, and dealing with practical situations like changing trains. His visual memory seemed to be intact; his challenges lay in making sense of what he saw. He had particular difficulty with faces, namely "the structured picture making up an individual, personal whole, which singles out every person unmistakably, and, in an anthropological sense, makes the person himself."[50]

S's face recognition deficit was at first difficult to notice due to the various compensatory skills that he developed. Bodamer designed specific tests to isolate S's challenges from his compensations, noting after these tests that "S recognized a face as such, i.e. as different from other things, but could not assign the face to its owner." He saw specific parts and was able to "identify all the features of a face, but all faces appeared equally 'sober' and 'tasteless' to him." He was also unable to make sense of expressions: he could note, for example, "anger or smiling, but he was unable to interpret them." He relied on external symbols like "hair or head covering" to "distinguish men and women," but "even then not always with certainty." (This stands to reason: hair and head covering are not now, nor were they then, a perfect or transparent indicator of gender.[51])

In the next section of his paper, Bodamer discussed S's own experience of seeing faces in both casual and experimental settings. First, "S is told to look at own face in mirror." He does not recognize himself and initially "he mistakes it for a picture but corrects himself." After "he stared for a long time as though a totally strange object is before him," he "reports that he sees a face and describes its individual features." He said that "he knows it is his own face but that he does not recognise it as his own." He recognized almost nothing about it at all, noting that "it could be that of another person, even that of a woman." When the test was repeated, S reported that "he is very distantly reminded of a similarity with himself," but, functionally, he had no relationship to his own face. He also did not recognize himself or his friends in a photograph, and he had to rely on "the help of clues from his clothing" when asked to pick himself out in the image.[52]

Further mirror testing asked S to "laugh and speak while others [around him] remained still." Same result: "He saw his movements but did not see his face as a whole, and did not recognise it as his own." Indeed, his own face "seemed that of a complete stranger." Time had no effect in increasing recognition: "After repeated experiments and observation over months, the patient's face grew no more familiar." Nor did the faces of family, which "were just as strange to him, in real life and in photos, as those of other people." He did not recognize his own mother when he came upon her by chance. He could never distinguish his fellow patients.[53] He did manage to differentiate between hospital staff, relying entirely on spectacles and uniforms; absent glasses or other external markers, for those in the same position and thus the same clothing, he was entirely at sea.[54]

S used similar techniques to identify figures in paintings; he pointed to "the cap" when asked how he recognized someone in a Dürer painting he had seen a few days prior. When he was again shown the painting with everything hidden but the face, he recognized no one and nothing. Similar experiments were conducted, all with the same results: "In no case was S in the position of being able to see in a face those things peculiar to the owner of that face, although he could distinguish physiognomic and expressive details astonishingly well."[55] He also retained an impressive visual memory from before his injury, and was able to "visualise faces, and describe them from his memory i.e from the time before the injury. . . . He asserted that he could imagine clearly and vividly the faces of relatives, army comrades and company leader, etc. just as well as he had ever done."[56] But he described a face he now saw as "strangely flat, white with emphatic dark eyes, as if made from a flat surface, like white, oval plates, all alike."[57] S also failed to recognize animals, as Bodamer recounts in experimental detail. Indeed, animals were even more challenging because S relied considerably and usually quite successfully on voice, gait, glasses, hair, and clothing to identify people; such cues were absent in both particular animals and animal species more broadly. He could identify none of these, confusing dogs and rabbits and people.[58] His challenges with faces absent context were particularly acute, however; while he could remember objects and scenes that he had already encountered, he was never able to recognize faces without other clues.[59]

While many of S's recognition challenges were constant, he also experienced periods of what he called "flickering" prior to which "everything turned almost white, like thick fog, like behind milk glass." Soon "a flickering came over both my eyes, then it was past." He briefly "saw quite normally for almost 10 minutes," though "the colours were not clear, but they were visible." During this respite "I saw well, almost as well as before my injury." Almost, "but faces . . . them I saw as badly as ever."[60] S used the language of sight to describe how he encountered faces; it was the best proxy for his experience despite the fact that the problem was not in his vision per se. The consistency of his face recognition deficit led Bodamer to theorize an independent pathway for face recognition in the occipital brain distinct from other recognition processes. In other words, prosopagnosia was not just one aspect of the agnosia whole, but a unique deficit connected to a unique part of the brain.[61]

Bodamer (a bit confusingly) presented Patient A as his second case. Patient A recorded that "apart from the eye I see the face blurred." In particular, "I don't see that which marks out a face." This extended to expression, as

"I don't see a particular expression of a face." Patient A sought out the eye first as "the most striking part of the face." After "I see the eye of a face, I go from there to the other parts of the face." Those other parts provided very little to A, as "when I look for that which is special about the face, I don't find it." This problem was hard for A, for whom faces held particular meaning and power. As Bodamer recounted, "Before his injury he had been a man who really saw faces." That's a striking statement in and of itself, making us pause to ask what it means to really see faces. Bodamer answered: "He had always been attracted by them, judged people solely according to them, and had also had excellent visual recall of faces."[62]

When A was injured, he lost not just the ability to recognize faces but to relate to others as he once had. He had to find an entirely new way to build relationships. This is perhaps true of many people; given A's avowed reliance on, and affinity toward, the face, he likely felt their absence all the more acutely. Because "I don't recognise people by their faces," he said, he relied on "the paraphernalia, not the facial structure." He averred that "I couldn't draw your face," and could not see whether a face is "sad or angry" because "the face means too little to me, the eye tells me more."[63]

Like S, Patient A failed to recognize people in images, even those he knew well. Of the book *The Sculptures of Naumburg Cathedral* whose pictures he'd looked at "many times before" such that "I knew them exactly," he lamented that "now I would not recognise them." This was to him very "sad" as "it's distorted." He did, however, recognize a picture of Hitler "immediately" due to the "moustache and the parting" of his hair. But "I'd not have recognized it by the face" as "the face itself I don't actually see, for me it's just a mass, everything is the same." This, to A, was not the way things should be, as "a face should really be differentiated and it's by the very differentness that you recognize a face."[64]

There's a kind of ache where recognition should be. When shown pictures of ten people, including himself, A "feels it to be familiar. He knows he's there." He doesn't know which one he was. Even though "he sees faces as different from one another," still "they remain a mass."[65]

The experiments continued: known people were swapped with unknown ones; two nurses with similar clothing and body types but "different in age and facial appearance" were indistinguishable.[66] Until "one smiles and he notices her unusually white and regular teeth." Only then did she become a person he could recall and identify; only then, in a way, did she become, to him, an individual person.[67] Like S, A could never recognize a portrait, no matter how often it had been seen. Like S, A relied on clues other than the face. And like S, A was drawn first and most powerfully to the eyes. The rest of the face remained a blur—no matter how hard he looked, or how long, or how often.[68]

Bodamer powerfully contrasted S and A to B, whose injuries distorted all faces such that "a nurse nose was turned sideways by several degrees, one eyebrow was higher than the other, the mouth was squint, and the hair shifted like an ill-fitting cap." B was in Picasso's world—for eight whole days, four weeks after his injury. And even still, "he recognized the nurse by her face, and never mixed her up with others." Faces were distorted, but still he recognized his mother by her picture on sight.[69] He did not have prosopagnosia, and even his "cerebral metamorphosia," which only affected his perception of faces, was temporary.[70]

The end of this essay included a summary of what Bodamer derived from these case studies and other historical examples. The cases, taken together, provide a long view of both face recognition challenges and approaches to and developments in neurology and the mind sciences. Many scholars have explored the intellectual history tracking the study of brains and their meaning.[71] Continuing to study these examples promises to be fruitful; while I will not delve too deeply into the results of such study, I'll note that we can see here how these doctors, scholars, and chroniclers made sense of face recognition deficits following disease or brain injuries. There are distinct traditions: Bodamer's emphasis on a distinct face recognition neurological pathway resonates with Quaglino, though his writings lie outside the canon and do not appear in Bodamer's work. Quaglino was strongly influenced by Gall's brain localization theories, without which the idea of a unique pathway for recognition would not make sense. Wigan and others were less committed to the unique pathway: for many of these scholars, recognition agnosia was connected to other agnosias. Bodamer insisted that it need not be.

In any event, there is a narrative here that took a long time—thousands of years!—to be granted a name, prosopagnosia, and systematized as a diagnosable condition. Even following Bodamer's clear-eyed and comprehensive analysis, it took another twenty years for a case to be clinically described, and then another nine after that.[72] There was another ten-year gap, and then slowly, the case counts began to build up.[73] Some would attribute this tremendous lag to the lack of technology that captured what was happening in the brain. That may be part of the story, but it ignores the technology that had already existed and was indeed being ignored: the technology of speech and of making sense of experience. When it came to listening (to mix a metaphor), the doctors were, indeed, blind. As we've discussed throughout this book, part of the difficulty in identifying face blindness as a thing was its lack of thingness. But it was also a problem of lived experience. Specifically, it was a problem of prioritizing lived experience as itself something to be studied.

Case 2 +: Hécaen and Angelergues (1962), Benton and Van Allen (1968), McConachie (1976)

And then, finally, face blindness became a thing; it became a syndrome. It took time, and the publication efforts were slow and steady, but following Bodamer's 1947 paper, prosopagnosia became firmly established as both an entity and an area of research. Hécaen and Angelergues published a paper on prosopagnosia in 1962, which was to date, as Benton and Van Allen noted six years later "the only study of a relatively large group of cases."[74] That failing was to change, but not for a while yet. Benton and Van Allen's own study was diagnostic and experimental, working with patients with cerebral disease to match both identical photographs of faces to one another as well as different photographs of the same face. This study marks a clear insertion of prosopagnosia into the research agenda; rather than describing cases, it seeks to identify them experimentally.[75] Helen McConachie published another single case study in 1976, referencing previous studies from 1967 and 1968.[76] I have highlighted three key papers that show increasing attention to prosopagnosia and its solidification as a category and, indeed, a problem. It took a name, a set of descriptions, a series of experimental approaches, and, as we have seen, increasing attention to the lived experience and stakes of the condition. Prosopagnosia was not an urgent research need then, nor perhaps is it now. But it was extant, and that existence, that coming into being, was quite a long—and at the same time, as we will see, quite a short—process, with roots in the nineteenth century and, equally, in the twenty-first.

Conclusion: Adding Up Anecdotes

I repeat that there is a great deal we cannot know from this collection of cases. We cannot say that they all represented the same neurological condition (not that such a statement would tell us much anyway). We cannot with certainty discuss what the brains of these men then tell us about face blindness now. (And why only men? There are reasons for this too: Whose experience matters? And whose deficits can be attributable to a break, an injury, a change, rather than always already having been wrong and thus unremarkable?) But there are also things we can and do know from thinking across these cases. We can know that living with face blindness, after one could once see, is hard. We can know that the patients experienced the loss of face recognition as, indeed, a loss. We can know that the doctors repeatedly tested their recognition not just of strangers but of loved ones, as if to emphasize that here lies the problem. As if to emphasize that losing the ability to recognize strangers is not really the point, but losing the ability to recognize one's caretaker, one's lover, one's mother, is a serious syndrome that must be demonstrated again and again to be believed. We can know, from these cases and testimonies, about empathy, about frustration, about recovery. We can know that people who cannot recognize others as they once did are different people than they once were. We can know—we can insist, guided by the field of health humanities—that while the exact pathways in the brain responsible for these lacks and losses matter, the lacks and losses matter even when the pathways are unknown.

And we can know that people lived with face blindness. People loved with face blindness. People developed compensations and strategies and ways of being in the world with face blindness. We can know that people once did this and that people still do. We can know, by thinking in cases, that the experience of deficit and disease and loss is both deeply individual and also collective, and that there is power in living with both. What may be true for many is not true for all but is useful for some. We can also honor the individual experience and learn from it and sit with its implication for the building of categories that help establish thingness. Which helps people understand and express what it is they might be experiencing. Their experiences are their own, but also there is comfort in the collective, as well as resources, and support, and knowledge, and power.

I have drawn here quite thin lines in the very mobile sand to excavate face-blindness-like examples in the past while insisting that I am not engaging in retroactive diagnoses. I argued that it's not just about finding lines of continuity but also differentiation. Surely it is valuable to compare how the inability to see or recognize faces was once described and understood with how we perceive it now. Surely by combing through cases in the past we can learn *something* about the trajectory of face recognition. We cannot directly connect what, for example, Thucydides once described (and I discuss in the coda) and what Oliver Sacks narrated millennia later, but there must be some ways to productively and interestingly think of these descriptions together. And apart.

The next chapter will reflect on face-blind physician and writer Oliver Sacks as a very special kind of narrator. We will consider his role, along with clinical developments and media narratives, in the modern invention of face blindness. Sacks was a clinician, and one who insisted, beautifully and with gentleness and force, on the nuances and therapeutic power of the particular. He was a diagnostician who thought in cases, not as a form of epistemology but as a form of empathy. He was a doctor. And he was, eventually, a patient. He told stories as both, and we have much to learn from both, and from the differences between them. Stories, like cases, create origins. Stories, about and alongside clinical research, create inventions.

2

The Blindness of Great Men; or,
How Prosopagnosia Was Invented

How do we go from the n of 1, or 2, or 3, or 6 known cases of face blindness across twenty-five hundred years to an n of 1–2 percent of the entire global population? How did face blindness become part of a spectrum upon which everyone falls? Or, to put it another way: Where were all the face-blind people before? They may have always been there, and we didn't know how to look. The diagnostic category of face blindness was not discovered but, like all such categories, created.[1] So a better question might be: What happened in the twenty-first century to create the face-recognition spectrum?

This is a historical process, and a cultural one, and, necessarily, a medical and diagnostic one. Face blindness is something that, for the second half of the twentieth century through the beginning of the twenty-first (and on), most people did not realize was a category. For an even longer time prior to the twentieth century, it simply wasn't a category at all. So we begin with the invention of the category. And I use the word invention advisedly: face blindness was not discovered, it did not emerge, it can't be isolated to a particular moment or event. It is a central claim of this book, and a great deal of all the scholarship in the history and sociology of medicine, that diagnoses are invented, as part of a process, in conjunction with bodies, cultures, technologies, people, places, and things.[2] Face blindness is one such cocreation, and a particularly interesting one, partly because it just took so long to build. And partly because, if we'd listened more to people's stories, if we'd even understood that this was something to talk about, it might have happened much sooner. Not that there is a standard timeline for creations, but when it comes to something as fundamental to human identity; modes of communication; the establishment and nurturing of relationships; and the nature of being in community with others as the recognition of faces, there was a wide scope for diagnostic invention and recognition. There still is.

The clinical category of acquired prosopagnosia (AP), which occurs following injury or illness, was a practical precondition for developmental prosopagnosia (DP), the sort that people are born with. We could imagine the second only with the establishment of the first. But even that came slowly and haltingly. As we have seen, AP was first clinically described by Joaquim Bodamer in 1947; it took fifteen years for Hécaen and Angelergues to collect a series of cases in 1962, with Benton and Van Allen reporting another case in 1968.[3] It was only in 1976—almost thirty years after Bodamer's initial AP report—that Helen R. McConachie published the first report of a case of developmental prosopagnosia.[4] We see a steady rise of research publications and case studies—slowly, with only a smattering of papers in the second half of the twentieth century, increasing in frequency in the early years of the twenty-first, with a sharp spike from 2012 onward. As late as 2003, an article on the state of research on face blindness noted that "due to the very low incidence of this syndrome, case reports are sparse."[5] But they were rising. Perhaps the creation of prosopagnosia is that rare but always imagined teleological story of scientists laboring away in the lab, getting ever better and clearer pictures of the brain alongside ever more sophisticated and widespread tests and measures.[6]

Of course, like all histories of science, technology, and medicine, face recognition is not the story of scientists, or not *just* the story of scientists. It is also story of—as so many things are—stories. In this case, stories told by one of the master chroniclers of the human experience. Prosopagnosia is in part the story of a storyteller doctor-turned-patient discovering his own experience as a category and helping to call it into being. Neurologist and writer Oliver Sacks crafted a new kind of medical career by paying attention to the nuances of the individual in the data of the many. Among his prolific works is a 2010 essay in the *New Yorker*.[7] Like some of his other writing, it's deeply personal, describing his own medical conditions, but unlike those works, he does not turn this one into a research project, a diagnostic mystery, or a reflection on doctor-patient relationships and empathy.[8] It's more of an admission, a kind of coming out, though not his official coming out, which happened even later, in his 2015 memoir, after a life lived mostly celibate.[9] In the *New Yorker* essay, Sacks wrote about his own face blindness, which he had lived with all his life but, like so many, came to understand only in middle age.

When Sacks wrote about something, people read about it. And learned about it. And some, possibly even 1–2 percent of the population (unlike so many other, far less common conditions that he chronicled) might recognize themselves in it. As we review Sacks's writing, we will go back in time to learn about the medical, institutional, and media structures upon which he was building as he learned that he was not just an occasional and sometimes temporary patient.

We will also consider the invention of a category—developmental and acquired prosopagnosia—and how it entered the diagnostic and documentary record as well as general consciousness. Prosopagnosia did in fact go from describing an n of 1 to describing 1–2 percent of the population. And the process both took a very long time, and happened very quickly, as we'll discover. The invention of face blindness as a category of knowable experience in the first decade of the twenty-first century (and beyond) made it recognizable, label-able, and applicable. It became a syndrome, a syndrome with a name, even without a clear singular crisis, a *before* of recognition and an *after* of lack. Instead, it gave shape and definition to a whole lifetime of crises that until then had been invisible. Oliver Sacks is part of the story as to why. So is the media more broadly, as is a shifting sense of what counts as diagnosable and whose conditions matter. And, of course, when you have an n of more than 0 and—despite the importance and value of the individual case— more than 1, something is not just an anecdote or a story or an exception: it's a research agenda. Or it could be, if enough people with enough privilege and access and status and organizational skill care to make it so.

Physician, Describe Thyself

You probably know who Oliver Sacks is. It's likely you've come across his work in one way or another. Did you see the 1990 film *Awakenings*, starring Robin Williams?[10] The film is based on Sacks's 1973 book of the same name, chronicling his transformative success in using the drug L-DOPA to temporarily awaken people with encephalitis lethargica, the so-called neurological "sleeping sickness" that leaves those affected motionless and speechless.[11] A prolific and moving writer, Sacks came to public attention with his rich writings that focused on empathetic studies of people with neurological disorders. He followed *Awakenings* with *A Leg to Stand On* in 1974.[12] The series of articles in 1985's *The Man Who Mistook His Wife for a Hat* became perhaps his best known written work, alongside his frequent contributions to the *New Yorker,* the *New York Times,* the *New York Review of Books,* and the *London Review of Books.*[13] By the time the film version of *Awakenings* was released, Sacks had become, in the words of literary critic Anatole Broyard, "a kind of poet laureate of contemporary medicine."[14]

There are a lot of hagiographies of Oliver Sacks.[15] There are also, of course, critiques, and they are powerful and resonant.[16] In reflecting on Sacks, artist and disability designer Sara Hendren wrote movingly about the fundamental challenges of writing about medical cases. How, she wondered, could she avoid musing on, in Jenny Diski's terms, "how interestingly wrong we can go?"[17] If Hendren wasn't quite ready to say, as disability theorist Tom Shakespeare did, that Sacks's work was a "high-brow freak show," and that Sacks was "the man who mistook his patients for a literary career" who "violates every principle of disability equity," she wasn't *not* ready to say that either.[18] Sacks's heroic narratives of patients overcoming disability (or not) play into a number of problematic tropes that undermine the lived experience of being disabled; at the same time, Sacks's unerring attention on the patient-as-person made space for representing disease and disability as part of—and only part of—the whole. He was trained, first and foremost, as a doctor. Sacks brought attention and voice and advocacy to autism and face blindness and a host of other neurological conditions that have often been misunderstood, excluded, and ignored. Legacies are complicated. People are complicated.

For literary critic Leonard Cassuto, Sacks was both empath and exploiter, at once a collaborator, a stagemaster, a cocreator, and a curator. Cassuto asked: "Was Oliver Sacks the P. T. Barnum of the postmodern world," collecting his patients' oddities and putting them on display precisely because of and for these differences?[19] Cassuto answered himself: yes, and. For Cassuto, Sacks's method was evolving and collaborative, working with his patients to not just tell but insist on their stories. This was important to them; this was okay with them. Shakespeare critiqued Sacks's rendering of autistic scientist and activist Temple Grandin; Grandin discussed how Oliver Sacks inspired her and changed her life.[20] People are complicated. Legacies are complicated.

Sacks was much beloved by both his patients and his audiences across print and screen. In addition to his compelling prose and gentle touch, Sacks's work was *interesting.* He revered the brain as "the most incredible thing in the universe" and saw his practice with it and journey to learn about it as one of life's great adventures.[21] And he was a man who had many adventures.[22]

Among the important neurological approaches that Sacks brought to light, his continued emphasis on the individual patient, the specific case, the lived experience of one person was particularly noted. Jenny Diski pointed out the difference between stories and case histories: stories have narratives and conclusions. Case histories don't. Sacks, she insisted, was writing stories, acting as author rather than doctor in his text. His audience was (mostly) not medical, and it was not meant to be. Much of his prose considered the after-the-fact, once he'd already found the answers, such as they were. His patients were on display *because* of their diseases and conditions. That's why they were there. That's what his audience was reading for. But while Sacks was writing in stories, he was a practitioner of what historian John Forrester called, as we have seen, "thinking in cases." For Forrester, this was the great innovation of psychoanalysis; for Sacks, this was the great responsibility of all clinicians and caregivers. The unit of analysis for Sacks was always one.

It wasn't just about empathy for Sacks, though that was a huge component of his practice. For Sacks, listening—truly listening—was treatment in and of itself, but also diagnosis: when it came to the weird and wonderful world of the quirky brain, Sacks was able to learn what he learned and know what he knew because he paid attention. A psychobiography reading might posit that Sacks was drawn to narratives and the word and the brain and listening precisely because he could not see faces or people, so he heard them instead. In his writing we see a longing to bridge worlds, even as we are always reminded that he was trained as a doctor and not a humanist of the lived experience. He described people; he listened to them; but he did not, it turns out, listen to himself, despite his long experience as an analysand. For much of his life, he discounted his own lived experience. He was the doctor, even when he was the patient. But he was an empathetic and caring doctor. People lined up to work with him, not just because he was, eventually, famous, but because when it came to the unexpected and unusual, Sacks was especially good at his particular and very personal job.

And it became more personal. In the middle of his career, Sacks-the-doctor became Sacks-the-patient, but not because he was getting older and sicker. That, as Sacks recounted in a *New York Times* article about his terminal cancer diagnosis, happened in 2015, as it will happen in some form or other to most of us who get older, and who all eventually die.[23] It's a precept of disability theory that we will all, if we live long enough, become disabled.[24] For Sacks, the disabilities happened sooner. Or rather, the disabilities were always there. It took him until middle age to name them and to experience them as such. But Sacks's neurological differences had always been there. And for a long time, he didn't even know. *Awakenings* aside, Sacks's public-facing methodology focused far more on description and diagnosis than treatment (and possible cure). We would not say to Sacks, "Physician, heal thyself," but we might have said, "Physician, describe thyself." It is striking that on the topic of face blindness, for the longest time, he did not.

Or is it? In failing to recognize his own face blindness until a trip to Australia to visit his brother in his sixties, Sacks, this narrator extraordinaire of the human condition, this detective of difference, this physician who among all others had the most finely tuned instrument of listening and seeing and then knowing, was just like everyone else. Sacks did not know he was face blind because he didn't recognize face blindness as a category. Almost no one did. Because as a widespread category, it wasn't invented yet.

That's the thing about keeping track and naming names: something never hits the documentary record if there is no crisis.[25] Without a crisis, there is no category. And if you could never recognize other people, then there is never a crisis. There's "just" life, in all its challenges and complexity. People born face blind never had a crisis. They never had a break or change. So that form of face blindness took longer to recognize. It's no surprise that acquired prosopagnosia—face blindness that people got after a traumatic brain injury or fever—came first. "Befores" and "afters" are quite useful for category creations. But that in and of itself was not enough. People were getting fevers and experiencing the consequences of infection for thousands of years before "face blindness" became a condition with a name. Some of these once fevered people could once recognize faces and, following the fever or other crisis, could not. Whether the condition was named, it was, experientially, known. The people who could never recognize faces at all had no crisis and had no record. Their experience has no name. Even that most practiced of observers of the human condition, even he who devoted a career to putting the unusual and uncanny and unnamed on display—to marvel, to appreciate, to gawk, to empathize, to study, to *know*—did not know this. Not in others, and not, for a long time, in himself.

And then Oliver Sacks became the story himself. What does it mean to go from being the one studying to being the one studied? Oliver Sacks—doctor, patient, storyteller, subject of stories—is a powerful way to enter this particular narrative that chronicles how something that may have always been there became a thing to be known; a case; a diagnosis; a condition; an identifiable way of being in the world. He played with this question almost obsessively in 1970's *Migraine* about a chronic but intermittent condition.[26] Sacks's account was both clinical and personal; it combined the neurological and the experiential meaning of migraines. As an account, it was possible because migraines as a phenomenon were already known; here, he was explaining and describing rather than discovering. Sacks returned to the theme of physician-as-patient in 1985's *A Leg to Stand On*, in which he explored what it meant to need—and only sometimes get—medical empathy from the position of the patient.[27] His leg condition—also, it turns out, a type of agnosia—was only temporary, and one he was able to tackle from a research perspective as well as an experiential one; this is a very different kind of patienthood. Prosopagnosia (so far as we know) is forever.

Sacks was one of many people with face blindness, and even one of many famous people with face blindness. His was not the first prosopagnosia case to be written about, but he may be the first one to write about himself, and doing so was consistent with his character. He wrote; therefore he was. He wrote about himself; therefore he was, in a way, someone new. Writing calls stories, and people, and, sometimes, categories, into being. And in so doing, Sacks helped prosopagnosia become not just an occurrence of a case or two, but a population-level condition. Even without a crisis. The unit for Sacks was always one, but he helped make himself one . . . of many.

The Dawning of Men

Great Men get enough ink. I don't want to make this the story of how Great Man of History and Medicine Oliver Sacks was responsible for spreading the word about prosopagnosia and in so doing, helped make it recognizable and thus widely meaningful—even though Sacks was indeed a Great Man of History and Medicine and did indeed play an important role in this narrative. We are going to go deeper than that. There's a version of this narrative where Oliver Sacks realizes later in life—to his bemusement and not-quite-surprise—that he, like his elder brother and possibly his mother, has always been face blind. Neither of them, like their mother, could recognize people; their n of 3 was in fact 3 cases in a much, much bigger cluster. His mother was certainly "almost pathologically shy." As a result of this shyness, "she had a small circle of intimates—family members and colleagues—and was very ill at ease in large gatherings." With the insight of hindsight, Sacks wrote that "I cannot help wondering, in retrospect, if some of her 'shyness' was due to a mild prosopagnosia."[28] Sacks, like his brother and maybe his mother, had a version of the condition that arguably was one for which he was famous, having chronicled the somewhat but not entirely related titular case in the bestselling *The Man Who Mistook His Wife for a Hat*.[29] He was unknowingly dealing with prosopagnosia (and struggling to deal with it) his entire life.

In fact, Sacks already knew about agnosias, in part from his own experience, as he chronicled in *A Leg to Stand On*.[30] And it took him until middle age to figure out that his challenges in recognizing people were clinical.[31] Without putting too much emphasis on this Great Man, that fact is worth pausing over. It's worth thinking about how conditions are invisible, unrecognizable, and unknowable, even by those *who already know them*, when the entire medical-social-cultural-diagnostic complex that fosters diagnoses is not, for a given experiential condition, in place.

But Sacks eventually figured it out. Surprisingly, given the detective-diagnostic genre at which he excelled, he didn't write in great detail about *how* he figured it out: in his prosopagnosia coming out piece in *The New Yorker* in 2010, he spent far more time describing what it was like (and what it had always been like) not to recognize others than describing what it felt like to learn that his and his brother's difficulty was not ineptness but inability. And not just any inability, but "something beyond normal variation, that we both had a specific trait, a so-called prosopagnosia, probably with a distinctive genetic basis." It "dawned on me" only when he visited his brother in Australia after thirty-five years "and discovered that he, too, had exactly the same difficulties recognizing faces and places."[32]

It *dawned* on him. This is quite a curious confession, and, of course, not entirely true. Sacks wrote in the *New Yorker* essay that after the publication of *The Man Who Mistook His Wife for a Hat*, he would get letters from people who saw themselves in his stories.[33] And from those letters (and *not* from his own stories), Sacks began to recognize himself. Finally. It took time, and many letters, and the experience of others, to make Sacks see. But it still did not quite "dawn" on him that he too had something, and that that something had a name, even though "I have had difficulty recognizing faces for as long as I can remember." This difficulty did not matter "too much . . . as a child, but by the time I was a teen-ager, in a new school, it was often a cause of embarrassment." Specifically, "my frequent inability to recognize schoolmates would cause bewilderment, and sometimes offense—it did not occur to them (why should it?) that I had a perceptual problem."[34] Why should it indeed?

For most people, a long-term lack of realization would not be particularly notable except insofar as they had heard about agnosias at all and then would have to self-advocate for identification. Actually, it would, in the first two decades of the twenty-first century, become rather a common narrative for the face blind: they would read about the condition in a newspaper, or hear about someone who had it, or come across a quiz online (Are you face blind? Click here to find out!) and learn that they could give a name to their challenges, and that what had been a source of embarrassment could, at least, be neurologically explained.[35] For many people, learning about their face blindness, or rather naming their face blindness, was a kind of dawning facilitated by the complex of media proliferation; increased medical awareness; anecdotal stories; and a broad coalescence of research that built on and contributed to all these other factors.

Sacks's dawning came on the heels of a concerted media strategy on the part of face blindness researchers in the early 2000s. The scientists at Harvard's Vision Lab were sick of schlepping across the country to chase after every possible lead, searching for any research subjects who might, just might, be face blind. If 1–2 percent of the population were face blind, Ken Nakayama and others theorized in 2006, surely there would be enough people in any given city, or state, or region, to test.[36] Surely there was no need to scour the country for something that, in the end, was not all that rare. Surely. But to find them, they had to know who they were. Nakayama's lab deliberately enrolled the media to support their efforts, talking at length not just about their research but about face blindness as a phenomenon itself, under the theory that if they publicize it, the face blind will come. In 2006, the *Boston Globe* published an article entitled "When Faces Have No Name," featuring prosopagnosia research at Harvard and around the world.[37] Other media outlets followed suit, and many face-blind readers (and also those who came to be known as super recognizers) of those newspaper stories and television programs recognized themselves (recognition—at last!) in those descriptions and narratives. For some, the recognition was instantaneous. For others, it was a kind of slow . . . dawning. And still others may not have thought about these pieces at all, but they clicked on a quiz. And they got a label. And that was mostly a big relief.[38]

The recognition of face blindness wasn't just due to Sacks, not by a long shot. It was a confluence of factors, but Sacks was a part of it, and a catalyst of it. Where Sacks once turned his own case into stories and descriptions (and research puzzles), here he did not. Sacks used his personal stories to distance himself from his own experiences by turning the case into a narrative. His story about his own face blindness was far more brief and far less clinical than his famous case studies, and in this way, it was perhaps far more personal.

What sense can we make of one person's story? What is the value of one case when we are trying to understand how one became 1 percent? Sacks himself was a great proponent of both the empathetic and the analytic value of the case; in part, we are following his lead in using his methodological approach upon him. But there is more, and it is not just that Sacks was among the most recognizable or high-profile face-blind people, though that matters. As Sacks himself discussed, his careful, detailed, and intricate observations of others that characterized his entire career were always missing a particular kind of description: a visual one. And his editors noticed. Sacks spent time with "memory artist" Franco Magnani in 1988, interviewing him and even traveling together for a profile in *The New Yorker*. When Sacks "finally submitted an article about him to *The New Yorker*, Robert Gottlieb, who was then the magazine's editor in chief, read the piece and said, 'Very nice, fascinating— but what does he *look* like? Can you add some description?'" For Sacks, this was an "awkward (and, to me, unanswerable) question" to which he responded "by saying, 'Who cares what he looks like? The piece is about his work.'" Gottlieb knew better, insisting that "'our readers will want to know,'" because "'They need to picture him.'" Sacks acknowledged the point, but was still genuinely unable to provide the needed description, so he turned to a trusted strategy of relying on his assistant, responding to his editor that "'I will have to ask Kate.'" Not knowing about Sacks's challenges with recognition, "Bob gave me a peculiar look." Kate, whom Sacks also could not recognize, helped Sacks out often, as it happens: "More than once, Kate has asked my guests to wear name tags."[39]

Of course, like his editor, Sacks barely understood these challenges himself, having always "assumed that I was just very bad at recognizing faces, as my friend Jonathan was very good at it—that this was within the limits of normal variation, and that he and I just stood at opposite ends of a spectrum."[40] Sacks wrote this as if to say he was wrong. But in a way, Sacks was right: he and Jonathan were indeed at opposite ends of, or at least quite far apart on, a spectrum. What a difference a name makes.

There's an irony in the fact that this person who was so deeply committed to description was also unable to describe himself. In fact, "on several occasions I have apologized for almost bumping into a large bearded man, only to realize that the large bearded man was myself in a mirror." It worked both ways: "Sitting at a sidewalk table, I turned toward the restaurant window and began grooming my beard, as I often do. I then realized that what I had taken to be my reflection was not grooming himself but looking at me oddly."[41]

That's what makes this such a good story—the ineffable magic in which the parts contributing to growing knowledge of face blindness add up to more than the whole. If even Oliver Sacks did not know he had face blindness, and wasn't just generally bad at recognizing faces, what hope could every other face-blind person in the world have of figuring it out? (Which raises the question: Does figuring it out even matter? We shall answer that question emphatically: yes. Yes, it does.)

A Butterfly Flaps Its Wings

Historically it has been hard for people with prosopagnosia to identify that their lack of face recognition is actually a condition with a name and research behind it. Most people just thought that others were much better at recognizing people. The publicity around prosopagnosia and the proliferation of diagnostic tests online led to a huge alleviation of pressure for many. Their inability had a name. It was real. They weren't just being rude or not trying hard enough. They weren't autistic, a diagnosis that was historically mistakenly assigned to face-blind people given their shared—but importantly different—struggle to make sense of some aspects of faces.[42] Sacks was himself "accused of 'absent-mindedness,'" and while "no doubt this is true," "I think that a significant part of what is variously called my 'shyness,' my 'reclusiveness,' my 'social ineptitude,' my 'eccentricity,' even my 'Asperger's syndrome,' is a consequence and a misinterpretation of my difficulty recognizing faces." In this, Sacks was one of many.[43] Of course there are some face-blind autistic people, and some autistic people who are super recognizers. But most face-blind people are not autistic, even if they for many years were told they were. There are a lot of ways to be bad at recognizing faces, and they do not all mean the same thing.[44]

We know that face blindness was clinically defined as prosopagnosia in 1947. But, as Heather Sellers wrote in her memoir about living with face blindness—or, more specifically, her memoir about her process of discovering that she has face blindness—most people who have it did not know what it was or could not identify themselves within it. Nor could their doctors. Sellers had to figure it out for herself and then convince them.[45] Dori Frame, who likely acquired prosopagnosia following an adolescent head injury, said in a *New York Times* health film that "when you put a name to it, it is the biggest relief of your life. It's kind of a weight lifted off you because you know you're not stupid or scatterbrained or not paying attention."[46] Frame had acquired prosopagnosia: she did have a crisis. And even still it was hard for her to get a diagnosis; like Sellers, she had to figure it out for herself. Frame told the *New York Times* in 2011 that she "didn't realize she had a problem until she learned about prosopagnosia in a psychology class." "'It's like colorblindness,' she said. 'You don't realize you see colors differently than anyone else until someone points it out to you.'"[47] Personal advocacy is a significant theme in face blindness narratives, but for even the most agile researchers, there needed to be a realization that this was something worth researching. You can't ask your doctor what you might have if you don't have the language to understand that you have something at all. It's no accident that Frame's story was told in the *New York Times* in 2011: after all, it was around then that face blindness was, in its way, invented.

In a *60 Minutes* interview in 2016, Sacks expressed why it was so hard to identify face blindness, saying that "It is not usually a complaint of people. People do not bring it up . . . One sort of assumes that other people are the way one is."[48] A host of compensatory mechanisms can make it even more challenging to identify, as Sacks wrote: "Many prosopagnosiacs recognize people by voice, posture, or gait; and, of course, context and expectation are paramount." While hugely useful, "such strategies, both conscious and unconscious, become so automatic that people with moderate prosopagnosia can remain unaware of how poor their facial recognition actually is, and are startled if it is revealed to them by testing."[49]

Mildly face-blind scholar and poet Jenny Edkins has emphasized the advantages of prosopagnosia, pointing out that a different way of accessing the world has its own benefits and carries within it a different politics. She offered a way of understanding prosopagnosia not as a lack but an opportunity to encounter people on different terms.[50] Emmeline May, known also as

Rockstar Dinosaur Pirate Princess, has written in depth about the particular skills she has that balance the challenges she faces as a severely face-blind person: "While the initial revelation [of my diagnosis] made me feel like everyone else had been walking around with a superpower (the power to recognise people) that I didn't have, I am also realising what the benefits are of my brain compensating for my lack of magical recognition ability. I can read facial expressions, micro-expressions, emotional states and body language LIKE A BOSS." These sometimes-eerie skills seem to access depths denied to those who are used to reading surfaces: "I often know how people are thinking or feeling before they even express it. (Which, I am told, is also really annoying sometimes.)" There are both personal and global benefits to a form of relationality in which facial features are irrelevant: "It maybe makes me poor at initial meetings, but it's helped me be a great communicator. It also helps me to be nicer to complete strangers. You never know, that person pissing you off on the tube might actually be the person who interviewed you for a job, so let's maybe NOT call them a shitweasel."[51]

Face blindness, these writers have suggested, has its own logic and valuable way of being in the world. Chapter 4 addresses why these advantages have not been studied fully, but it is clear that in addition to the challenges that face blindness presents—and there are many, and they have a whole host of implications—there are not just compensations but alternatives to sociality based on face recognition. Face-blind people tend to have highly diverse communities, forming relationships with people with distinctive features, a range of racial backgrounds, and differing heights, weights, and gender presentations. It makes them easier to identify. Also, face-blind people aren't drawn to traditional markers of desirability and attractiveness, because those markers are meaningless signifiers to them. (Though of course attractiveness can have a significant impact in the way people are treated and highlights the nuances of what appearance can mean and the effects it can have.[52]) Such adaptations also can present barriers to realizing that face blindness is a category; face recognition is so ubiquitous and so necessary that people without it have a lot of workarounds.

The 2006 *Boston Globe* article discussed parents who could not recognize their children if their clothing was changed at day care; clothing cues, like hairstyles, unusual piercings, and glasses, are a powerful but limited adaptation, because they change. Face-blind parents had to contend with the staff's concern with parents who did not recognize their children, especially when those parents could not themselves explain why.[53] Without a diagnosis, and without people talking about face recognition as a category, and without widespread awareness of the condition, how could these parents know? In the 60 Minutes segment with Sacks, fellow face-blind interviewee Ben Dubrovsky chimed in that "it never, ever in my life occurred to me that people would look at a face and just get it like that." Jo Livingstone echoed these sentiments and underscored the personal and social stakes in being unable to do what so many others found easy: "I believed that I was not good with people but I had no idea of the reason. I just thought I was stupid." It was only when Livingstone read the Globe article about prosopagnosia that she presented herself to featured face blindness researcher and neuropsychologist Brad Duchaine's lab for testing, and, indeed, confirmation of her condition.[54] Wendy Holt also turned to Duchaine after "finding out about Dr. Duchaine on the internet."[55] Sometimes even the medical doctors don't know; but in this case, a few of the scientists do.

And getting that label, that name, can be an enormous relief, as Sacks underlined: "People do think you may be snubbing them or stupid, or mad, or inattentive. That's why it's so important to recognize what one has. And to admit it."[56] Wendy Holt shared this perspective, saying that "when I was diagnosed, and it was clear we had this inherited condition, I cried—out of relief for understanding what was wrong, and also out of shock." The label brought not just understanding but practical help: "Now I know there's a specific reason for my behaviour. I'm far more on my guard against doing anything that makes me look foolish or appear rude."[57]

For both Heather Sellers and Jo Livingstone, getting the label meant they could share it with others. Livingstone now "always explains to people she meets that she has a problem with her memory and that if she sees them again she may not recognise them."[58] Sellers's diagnosis provided an explanation for both the present and the future, helping others understand why she couldn't recognize them and—importantly—making space for building a different kind of relationship. One that wasn't always undergirded by the fear of ignoring a loved one and the anger of the loved one who was ignored.[59]

Dori Frame likewise reflected on one of the practical social implications of recognizing no one, ever. She said that "I always enjoy being around strangers. And I notice I like to move a lot, I like to travel, I like to be in an environment where I don't know anyone. That's when I'm most comfortable is when

I'm in a sea of strangers."[60] Because she is always in a sea of strangers, a world of blank faces tells her nothing about her relationship to them. If she knows for sure they are indeed strangers, she doesn't have to worry about failing to recognize someone, the source of her significant social anxiety, common to people with face blindness, a category she can now name and share.

Should we call the recognition of face blindness "the Sacks effect"? When Sacks started writing, people started realizing. Because, like him, they did not know what had been there all along. And in a way, even though they had always been face blind, until this rush of interest and attention and information, perhaps face blindness itself had not been there all along. By some measures, including the experiences of those who have it and their ability to identify it, it was only in the first two decades of the twenty-first century that face blindness was truly invented, despite being named over fifty years earlier. People did not realize they were face blind because they could not until face blindness was invented. So we are back, in unexpected ways, to Great Men of Science and Literature. It wasn't just that a prominent neurologist was writing about this condition in public forums so soon after the concerted media strategy launched by face blindness researchers, and it wasn't just that a well-known writer was discussing his own experiences with being face blind. It was the particular confluence of the well-known writer of interesting and quirky cases about the brain who is also a neurologist who was *also* himself face blind that captured the public imagination.

It gets better. There's a kind of buddy comedy here, but I hesitate to stretch the bro metaphor too far, given the sexual misconduct allegations against the other side of the duo, face-blind portrait artist Chuck Close (1940–2021).[61] Close, like Sacks, seems to be an unlikely candidate for face blindness. A premier portrait artist, Close is known for his giant canvases that realistically reproduce faces from afar, while, when examined closely, show pixilated segments whose parts have their own logic and visual interest.[62] He made big faces from small pieces, and those small pieces have a life of their own. Close described himself as a "rolling neurological clinic"; in addition to face blindness, he had dyslexia (discovered only when his daughter was diagnosed with it in first grade), muscle weakness, and, following a spinal stroke in 1988, was largely paralyzed.[63] From a young age, Close used his artistic skills in school and other contexts to demonstrate that, in his words, he "wasn't a slacker."[64] Like Sacks, Close's face blindness remained undiagnosed until much later in life, though he always knew that, among his other challenges, he could never remember faces.

Close was clear that he became a portrait artist not in spite of his face blindness but because of it. In his own particular form of therapy, Close used his portraits as an unconventional—but for him, necessary—way to commit faces to memory . . . and posterity. Close was unable to recognize three-dimensional faces, but he had no such issues in two dimensions. Flatten the face (as he did in his mind's eye and on paper), and it was no longer, for Close's face blindness, a face as such—and then he could recognize it. The process of flattening a face and then literally rendering it helped him remember it. Close suggested a fascinating shift in how the face is processed, turning it from one cognitive mechanism that, in Close's case, didn't work as well, to another, which did. It was still a temporary fix—Close would recognize the two dimensional flattened face only if he was looking at it straight on, so once someone turned slightly, or left, he had to do it all anew—but one that produced images that have been recognized by inclusion in the Museum of Modern Art and the Whitney Biennial, and even by Close's receiving the 2000 National Medal of Arts. For Close, painting faces was intimately intertwined with being unable to recognize them; portraiture is both a function of his face blindness and its treatment. He began to draw faces because he couldn't see them. And then he (and the world) could.

In addition to the neurologist who never knew he had a neurological condition, we have the portrait painter who couldn't recognize faces. Both were creating pictures of humanity, Close in images and Sacks in language. On the World Science Festival stage in 2010, Close said to Sacks, "I always saw your narratives as art."[65] Building on the momentum of Sacks's article, Sacks and Close appeared together at venues like the World Science Festival to allow the public to marvel at this odd coincidence of factors. Which, of course, had the benefit of letting people know not just about these two face-blind people, but face blindness itself. The story was such a good one that it was perhaps inevitable that these two face-blind artists would take their show on the road. Their particular life histories were tailor-made for pithy headlines and catchy ledes. Numerous newspaper articles and reports told the story of face blindness with a direct focus on Sacks and Close, drawing on the narrative tradition of the eccentric artist and the eccentric genius. Sacks did it himself in his 2010 *New Yorker* article, writing about how Close's prosopagnosia "played a crucial role in driving his unique artistic vision." Close told him that "I don't know who anyone is and essentially have no memory at all for people in real space . . . But when I flatten them out in a photograph I can commit that image to memory."[66]

The buddy show started around 2010. But that was already kind of late in the game; the Prosopagnosia Research Center had formed in 2001 and had already developed a rich research agenda with ever-growing numbers of participants and subjects. Harvard postdoctoral fellow Richard Russell and his team published a game-changing paper in 2009 positing the existence of super recognition, itself a new kind of diagnostic label, and notable also for upending the way that researchers had hitherto understood the entire face recognition process and its deficiencies. With the categorization of face blindness came the reorientation of face recognition as a spectrum, exactly as Sacks put it. Prosopagnosia, then, was not a pathology, a lack, that people either had or didn't, to varying degrees. It was one end of a spectrum; most people in the world fall somewhere in the middle, as is the way of spectrums by design, and some were at the other end. The most face blind were at one extreme, and the super recognizers were at the other. This is another piece of the puzzle; it was Russell's group that developed the media strategy that made face blindness more well known. And, of course, like all things twenty-first century, we must consider the role of the internet.

Untangling the Web

I could claim that the key factor in growing attention around prosopagnosia was the now easy availability of online quizzes. You know the genre; maybe you've taken one or two yourself. Want to know which *Gossip Girl* you are? What *Hunger Games* district you'd be in? Are you an introvert or an extrovert? There's a quiz for that. And there always *were* quizzes for that; the entire teen magazine industry was built on them.[67] But, with the internet, you didn't need someone else to score your answers; you could do it yourself with a few clicks of the button. The results would be generated for you for free (in exchange for your data) and instead of laughing about it with a friend, you could ponder the results yourself. It's maybe a little lonelier to learn your best shade of lipstick or what kind of mythical beast you might be on your own, but it also provides space to quietly reflect, ponder, and learn. Especially when it comes to diagnoses.

Google's medical credentials can be a little dicey. I don't want to overstate the role that online face blindness diagnostic tests played, nor do I want to place too much emphasis on either the internet or newspaper publicity as the drivers of prosopagnosia awareness. Each of these factors alone, alongside Sacks's high-profile writing and storytelling, is part of this puzzle. And we can get more granular about both *when* these pieces played their roles, and what kinds of roles they played. While these things are complicated and interconnected and dependent on multiple social, political, economic, and cultural forces, we can, with confidence, again note that the n of 1 or 2 became 1–2 percent. Of the entire population—which is a lot more than 1 or 2 or 3 or 4.

To be clear: the research had already started. The Prosopagnosia Research Center across Harvard, Dartmouth, and University College London was founded in 2001, building on a slow but steady stream of face blindness research by a number of scholars including Harvard Vision Lab founder Ken Nakayama. Dartmouth professor Brad Duchaine, then a postdoctoral fellow in the lab, built the faceblind.org website linking the three universities and their research; this site, alongside others including Sarah Bate's prosopagnosiaresearch.org at Bournemouth University and the face recognition research site at the University of Washington, remains a key way that information about face blindness is accessed and proliferated.[68] Most of the scientific literature situates the growth of prosopagnosia research as beginning in the early 2000s; Murray and Bate wrote in a 2020 article that "over the past 20 years, research into DP has surged," echoing Arizpe, Saad, Douglas, Germaine, Wilmer, and DeGutis's 2019 contention that "over the last 20 years, interest has been growing in identifying and studying individuals with poor face recognition abilities, especially those with developmental prosopagnosia."[69] These are non-empirical claims, but they are borne out by the publication of scientific papers, the proliferation of research centers, and, indeed, rising newspaper and screen coverage of the condition.

And the quiz.

The first online face blindness quiz was mentioned in the 2006 article in the *Boston Globe* that played a pivotal role in opening up a public discussion about face blindness. The article includes a link (such as it was in 2006; most people were still not accessing their newspapers online at that point) to the short form of the Cambridge Face Memory Test (CFMT), with the caveat that "it is not publicly available because researchers do not want potential subjects to become familiar with it." For those who were interested enough to seek it out, "Duchaine offered a link for readers who want to quiz themselves: www .icn.ucl.ac.uk/ facetests/."[70] The original 2006 CFMT was developed by Brad Duchaine and Ken Nakayama, improving upon prior face recognition detection mechanisms.[71] The short CFMT and its longer form, the CFMT+, were the gold standard until quite recently.[72] According to Brad Duchaine, then at University College London, his lab manager, Laura Germaine, "created an alternative CFMT and made it available in 2005 or 2006"; the alternative form protected the integrity of the original quiz and drew over fifty thousand users when it was originally made public.[73]

The quiz is not a direct proxy for the longer test, but it did give people results they could follow up on to get more information. And it does work along similar principles and even uses some of the same images, which is why the Vision Lab was initially so protective of it and tried to limit access to it. Taking the test served multiple purposes beyond diagnosis; regardless of how people scored or even why they were taking it, they now knew face blindness existed. While some clicked through because they suspected they were encompassed by the category of face blindness, others were just intrigued (aren't you?). Many people came to the quiz through newspaper articles and television shows; others sought it out specifically as they came to learn about face blindness from friends and other sources. The internet in the early 2000s was a very different place than it is today, but it was still a destination for people to learn more about what piqued their interest. And face blindness piques interest for those who don't have it because it is so hard to fathom. And it piques interest for those who do—because for them, it isn't.

In addition to the modified version of the CFMT, there are sites that offer a series of questions that prompt users to think about their own experience of face recognition. (Do you have trouble recognizing faces? Would you say you struggle to remember people after meeting them?) These are not face blindness tests per se but rather provocations that direct users to understand that face recognition is a category in and of itself. For those whose answers indicate poor face recognition, the site offers further information on prosopagnosia and directs users to contact various researchers in the field for follow-up testing. These sites began to proliferate around 2006 and onward, consistent with the rise in research and testing access.

The test isn't perfect. Tests rarely are. The test itself reproduces many biases that can affect the results: the initial version featured white computer-generated faces; research has overwhelmingly shown that people recognize those of their own race far more effectively than other races.[74] Another popular diagnostic asks users to identify faces of celebrities, which is slightly different than the "before they were famous" test that is deployed to test for super recognition. While failure to recognize, say, Madonna or Barack Obama could certainly speak to poor face recognition, it could also be indexed to a different cultural framework or access to news and popular culture. These tests *may* indicate face blindness, but they could also indicate something else entirely. The full testing regime is rigorous and multivalent, but these snapshots are a clear indication that testing itself can reproduce assumptions and biases that make the results more complicated. Someone who fails to recognize Barack Obama has failed to recognize Barack Obama. That person may fail to recognize everyone, or they may just fail to recognize recent US presidents, or even people of other races and ethnicities. Again, the diagnostic regimes conducted in the research labs often account for such issues, but the online tests do not. Because tests are not perfect. Nor is diagnosis. Indeed the recently released Oxford Face Matching Test was developed in part to account for the ways that autistic people and face-blind people assess faces—in parts, focusing on specific features—which is different from the holistic processing that scientists suspect most other people do.[75] The CFMT, the authors of this new test claim, tested for failures of holistic processing; poor results demonstrated piecemeal processing that was a feature of, but not unique to, prosopagnosia.[76] Again: there is more than one way of being bad at recognizing faces, but autism and face blindness are not the same.

A recent essay explored the accuracy of self-diagnosis in face recognition deficits and found that while many face-blind people reported themselves to be in the average or even high range of ability, self-reporting is highly valuable as both a diagnostic and experiential mechanism, particularly in the first instance.[77] The authors considered face blindness alongside dyslexia as a kind of experiential category. Both, they posited, might be something people could figure out about themselves as long as they had context and language to understand that what might feel like a lack of attention or skill or smarts is, in fact, a diagnosis. And the very fact of a diagnosis changes the tenor of the experiences.

That's a very current kind of discussion, possible only because face blindness is a known entity; in the early 2000s, the question was not how good people might have been at identifying the severity of their own face blindness, but rather that it was a category at all. Following the publication of the 2006 *Boston Globe* article on face blindness research, there was a modest uptick in media interest in the condition: several additional articles were published between 2006 and 2013, including pieces in *Wired*, the *New York Times*, and *National Geographic*, and features on *ABC News* and *60 Minutes*.[78] The rise in research attention anchored by the formation of the Prosopagnosia Research Center in 2001 and followed by increased publication activity, alongside the dissemination of online face blindness quizzes, a proliferation of newspaper articles and media attention, and public discussion of face blindness by high profile people like Oliver Sacks and Chuck Close, put face blindness on the map. By 2013 there was a flurry of memoirs, notable public diagnoses, and rising awareness of face blindness.[79] While prosopagnosia had been clinically defined by the second half of the twentieth century, and identified well before that, it was only in the first decade of the twenty-first century that it was, in any meaningful way, invented.[80]

There are a lot of empirical-ish ways to prove that prosopagnosia only became a major focus of research attention in the early 2000s, and only became a focus of somewhat widespread awareness in the first two decades of the twenty-first century. I've done them: the Google ngrams and interest graphs and internet archive and media searches. I've looked at the rise in the use of, and searches for, the term "face blindness" and "prosopagnosia." I've produced a lot of data that could feel robust and I could display it here, which might in a way make this feel more convincing. But I'm a bit suspicious that these kinds of numbers don't mean a great deal on their own, and that they reproduce the question rather than offer the answer.[81] I've tried instead to trace the underlying developments that led to these spikes and to explain some of the broader contexts in which they occurred. The system of face recognition was made visible through moments of failure, increased attention and discussion, and, simply, listening to people's own experiences.

Buffy Made High School Actual Hell

Despite the heading, this section has nothing to do with *Buffy the Vampire Slayer*, important as it was to me at a certain point in my life. (I didn't know about Joss Whedon's pattern of assault and mistreatment of women then, though honestly, at least some of it should have been clear.[82]) The brilliance of *Buffy* (or at least one of the many brilliant things about *Buffy*) was that it took the metaphor of high school as hell and made it literal. The series of Big Bads was a perfect analogy for the experience of so many in that moment, making this show about vampire slayers deeply relatable.[83] High school and young adulthood is for many a time of deep questioning, personal transformation, and the search for meaning and identity. And what *Buffy* did for high school demons, face blindness has begun to do for high school identity. What could be a better way to frame the young adult's often painful search for identity than a category of young adults for whom the very seat of identity—the face itself—cannot be recognized? Which is as clear a way as any to signal that, as a cultural category, face blindness has arrived.

That's still a bit opaque. More specifically, face blindness has emerged as either a feature or a key plot point in YA (young adult) fiction, playing on precisely the same dynamic that made *Buffy the Vampire Slayer* so compelling (and meaningful, and, for many, necessary). And it was set in high school, when it is already hard to figure out everyone's identity, including one's own. It's even harder if you literally cannot tell people apart. Face blindness is a powerful lens through which to narrate the teen experience, as evidenced in Laura Ruby's 2015 *Bone Gap* and Jennifer Niven's 2016 *Holding Up the Universe*.[84] These coming-of-age novels explore the story of love, betrayal, and identity at a time when identity itself is constantly being explored and altered, through the eyes of people who cannot see identity as writ on the face. As *Buffy the Vampire Slayer* made high school actual hell, face blindness novels make high school identity actually blind.

The timing is not an accident. It takes a while to write a book. For Ruby and Niven to have used prosopagnosia as a key feature in their novels, they had to know it existed in the first place. And it had to exist in the first place. Call it stage 2 in the invention of a diagnosis: first it has to be made and known—that's stage 1, and it's a complicated and interconnected process, as we've seen—and only then can it be applied not just in medical but cultural contexts. Note: sometimes the application can come before the medical reality; many face transplant films were made well before the first face transplant operation in 2005.[85] But that only supports the nature of this process: even as face transplants were not a medical reality, they were a conceptual one. People knew they were a potential thing even before they were an actual one.

It isn't subtle. The blurb for *Bone Gap* reads:

Everyone knows Bone Gap is full of gaps.

So when young, beautiful Roza went missing, the people of Bone Gap weren't surprised. But Finn knows what really happened to Roza. He knows she was kidnapped by a dangerous man whose face he cannot remember.

As we follow the stories of Finn, Roza, and the people of Bone Gap, acclaimed author Laura Ruby weaves a tale of the ways in which the face the world sees is never the sum of who we are.[86]

Finn cannot recall or recognize the face of the kidnapper because he has prosopagnosia. As the novel progresses and (orphaned) Finn discovers love and all those other hallmarks of YA fiction, he sorts through what it means to see, to recognize, and to feel empathy and connection. As do those around him, against the backdrop of midwestern cornfields and the magical realism of shifting worlds. Face blindness is one of many plot points, but it makes the question of identity and recognition all the more urgent, opening up new ways of knowing others (and oneself) and offering face blindness as a form of difference that complements the ways that everyone is unique.

Sure, it's a little schmaltzy—it's a back blurb. I could spend a lot of time here analyzing the novel, and I would enjoy it. But, unfortunately, it's not a key part of the argument I want to make here, which is that in the second half of the twenty-first century, face blindness arrived. When something hits YA novels as a way to make sense of the senseless experience of high school, it is officially a Thing.

Holding Up the Universe came out a year after *Bone Gap*. Where *Bone Gap* is a deeply literary intervention, drawing on magical realism, stirring language, and creative and experimental devices to tell its story, *Holding Up the Universe* is a more straightforward genre piece. It's also a love story, but without a lot of the extras. In *Holding Up the Universe*, high school is (as in *Buffy*) hell. Not being able to recognize people makes that hell worse. Especially as it is a deeply guarded secret, as per the jacket copy. Ahem:

Everyone thinks they know Libby Strout, the girl once dubbed "America's Fattest Teen." But no one's taken the time to look past her weight to get to know who she really is. Following her mom's death, she's been picking up the pieces in the privacy of her home, dealing with her heartbroken father and her own grief.

Now, Libby's ready: for high school, for new friends, for love, and for EVERY POS-SIBILITY LIFE HAS TO OFFER. *In that moment, I know the part I want to play here at MVB High. I want to be the girl who can do anything.*

Everyone thinks they know Jack Masselin, too. Yes, he's got swagger, but he's also mastered the impossible art of giving people what they want, of fitting in. What no one knows is that Jack has a newly acquired secret: he can't recognize faces. Even his own brothers are strangers to him. He's the guy who can re-engineer and rebuild anything in new and bad-ass ways, but he can't understand what's going on with the inner workings of his brain. So he tells himself to play it cool: *Be charming. Be hilarious. Don't get too close to anyone.*

Until he meets Libby. When the two get tangled up in a cruel high school game—which lands them in group counseling and community service—Libby and Jack are both pissed, and then surprised. Because the more time they spend together, the less alone they feel. . . . *Because sometimes when you meet someone, it changes the world, theirs and yours.*

Jennifer Niven delivers another poignant, exhilarating love story about finding that person who sees you for who you are—and seeing them right back.[87]

Once again, face blindness is a key plot point, embedded among other classic YA hallmarks including parental loss, body insecurity, high school trauma, identity secrets, and finding unexpected love. In both novels, the worlds of the protagonists change; in the case of *Bone Gap*, they change quite literally, given the magical realism framing. And in both novels, coming to terms with a hidden aspect of one's identity—prosopagnosia—helps the characters make sense not just of the identity of others, but also themselves. That doesn't mean high school is necessarily any less hellish, or that identity is any less opaque, but for these characters, there is the opportunity to grow, and change, and not just inhabit new worlds but help create them. Face blindness is both metaphor and mechanism, a way to make concrete the challenges of recognizing others while also, eventually, setting the stage for acceptance.

Incidentally, these novels also raise awareness about face blindness. If you missed the quizzes and the newspaper articles and the *New Yorker* story and the memoirs and the tv specials, you might catch the details in a YA novel. And, if you are a young adult (or old adult) who can't recognize faces, you might, in seeing yourself in others, feel just a bit better. And you might also search for prosopagnosia and take a quiz and get a preliminary diagnosis to take to your health care practitioner. And then you too will become an official part of the 1–2 percent of the population projected to be somewhat or completely face blind. While there are adaptations and compensations, there is no cure, no effective therapy, no long-term improvements or changes.

Which isn't to say that identification of the condition doesn't matter to those who are diagnosed: it does, hugely, helping the face blind to change their self-image and to understand the challenges they've faced. Naming matters, and it alters things in the world: the interaction of the person and the label fundamentally changes both sides and can create something that did not exist before.[88] The label of "blindness" calls a whole set of powerful associations into being; so too, as we'll see, does the word "super," and indeed, so does the concept of a spectrum. With the 2009 introduction of the supers, the face blind get recategorized as no longer pathological or damaged, but remain a medically motivated and research-created identity category.[89] Once diagnosed, prosopagnosiacs and super recognizers occupy ambiguous categories of medical affinity and biosociality: at one extreme, there is no treatment or cure, and at the other, there is very little obvious application. But the naming alone, and the ritual of testing, establishes a category and reinforces a kind of "patient brain" that helps call identity into being.[90]

Conclusion: Where It Started . . . How It Ends?

And *that's* how you go from an n of 1 or 2 to an n of 1–2 percent. You invent a category that encompasses a population. Or at least, the rise in media attention, coupled with easily accessible tests, growing research interest and activity, high-profile diagnoses and discussions, and increasing cultural penetration, is part of the story. Prosopagnosia research and newspaper publication are part of the story. Oliver Sacks and Chuck Close are part of the story. Heather Sellers and Dori Frame are part of the story. Television shows, interviews, and novels are all part of this story of naming and its implications. But only part. What about the other 98 percent? And if there is a bottom 1–2 percent, is there a top? A category has been invented, but it lies on a spectrum. When did that spectrum come to be? In chapter 3 we move to the other end of the spectrum, exploring the invention of super recognition as both its own category and a way to reconfigure face recognition as a whole. We'll meet three key players, Richard Russell, Josh Davis, and Jennifer Jarett, who generously entrusted me with their stories, thereby introducing a fourth key player into the mobilization of the super recognition narrative: me. We'll explore their histories alongside the scientific literature, and, later, the experience of other super recognizers to both see the other side of the spectrum that instantiated the spectrum itself, and think about how the creation of history and the writing of history and the construction of history mirror the creation and writing and construction of categories and the creation and writing and construction of diagnoses, and, indeed, the creation and writing and construction of the face recognition spectrum. Or, to put it another way, how we go from an n of 1 or 2 or 1 or 2 percent to an entire population.

3

More Men, More Invention

The Other Side of the Spectrum

(and Two Sides of the Same Story)

It was a career-making discovery. Too bad it wasn't the career that Richard Russell wanted.

As a postdoctoral fellow at Harvard's Vison Lab, Russell was part of the team that first clinically identified super recognizers as part of their prosopagnosia research. He was first author on the (rather rushed) academic essay that presented the condition.[1] He named it (reluctantly). He spoke about it (publicly). He worked on it (extensively). And then Richard Russell, poised to have a high-powered public-facing academic career that looks a lot like that of University of Greenwich professor Josh Davis, walked away. If there is an origin story to the creation of super recognition, Russell is right at the source. And at the end of his postdoc, he accepted a position at a low-profile, small liberal arts college, following his research interests and dreams, which were decidedly not about super recognition. In today's desperate and cutthroat academic environment, one in which research about face recognition (and its biases) is garnering multimillion dollar grants from royal foundations and MacArthur "Genius" Fellowships, it seems a puzzling choice.[2] Was Russell walking away from a guaranteed future, or toward a better one? It depends on whom you asked. I asked Russell himself. This is his story.

But of course it is not only his story. Once I became interested in the face recognition spectrum, and especially once I talked to Russell and others, it became my story as well. And once Russell and his team began to publicize their research and goals, it became a widely shared story, which began to change it. And even as it was Russell's story, it was always also the story of the people who lived these experiences, even before Russell named them. But even that naming was contingent upon Russell encountering his research subjects, and it is very much their story as well. A few years after I first spoke to Russell, I was able to interview Jennifer Jarett, the first clinically identified super recognizer. This is her story as well, and, as we'll see, it sounds a lot like Russell's. They came to this together, in a way, and it makes sense that they share a narrative. But Jarett's lived experience gives her a different perspective, as do her own goals, frameworks, and ideologies. As do everyone's. We'll see examples of these differences with another super recognition researcher, British academic Josh Davis. Davis's context is different, as are his institutional commitments and constraints. In a way, Davis picked up where Russell quite deliberately left off, helping to turn super recognition research into a professionalized endeavor. And also a capitalist one, and a criminological one. It is from all of these that Russell walked away.

We have learned the general story of face blindness. We've seen both the big picture of diagnostics and dissemination and the establishment of categories, and the individual narratives of cases and how they fit together. As with all things to do with the body and all things to do with categories (and all things to do with . . . people), face recognition is both general and highly specific. So too is the practice of history. By writing this book and interacting with this material, I myself have entered the story. And my own networks come into play in meaningful ways: I sent Richard Russell an email in the fall of 2018, and he generously agreed to a phone conversation, which we had that October. I also reached out to Josh Davis, who kindly spoke with me a number of times over the years. And as my friends and colleagues learned about this project, they began to share anecdotes about the people they know who might also be part of the story.

One name kept coming up, both in my research and through my networks: Jennifer Jarett, the first known super recognizer. I assumed she would be hard to contact and busy, as evidenced by her various media appearances in high-profile venues like the *New York Times* and *60 Minutes*. I should have just reached out early on, but I was, I admit, a bit intimidated. It turns out, though, that Jarett went to high school with two of my friends in the greater Philadelphia area; using their names made it easier. And it turns out that Richard Russell's in-laws also live in Pennsylvania, which made some of his connection to Jarett easier. We tell stories of science and research and how they progress based on findings, and ideas, and material, and even funding and happenstance and luck. Science and technology scholars are often attentive to networks and connections and fortuitousness in the way that research develops and travels; here, I want to be reflexive about my own role in accessing and indeed creating such a network. And in creating and disseminating research: this book too is a part of the story of face recognition, and I too play a role in it. This book now takes a place in the official history of face recognition.

I spoke to Richard Russell, Josh Davis, and Jennifer Jarett. They have different things to say about super recognizers, and what they can do, and what they should (and should not) do. We'll see the invention of a new category: the framing; the hype; the applications; the backlash; the next steps. We'll hear about plans for the future, which continues to change alongside face recognition technology. We'll see the people operating in lieu of, and then in conjunction with, the machines. We'll note overlaps between Richard Russell and Jennifer Jarett's stories, and the points of divergence. There's a difference between studying something and living it, even if the one studying it gives a name and context for the one living it. And Josh Davis applied it. Richard Russell walked away. Josh Davis is trying to build a super recognizer empire. And while supers around the world are trying to figure out how to use, apply, and monetize their skills, Jennifer Jarett just wants to live her life.

Inventing a Superhero

Super recognizers (the name isn't great, but it was the best Russell came up with) are just that: people who can recognize others much, much better than the rest of us. Most people remember about 20 percent of the faces they see; supers clock in at over 80 percent.[3] It works in a couple of ways: super recognizers remember every face they've encountered, even fleetingly, so that if they again see a face they once passed in a crowd, or meet someone they served (or who served them) in a restaurant or store, or finally get introduced to that person who is clearly on the same bus commute, they will recall their features. They will know that they have seen that person before, and they will likely be able to situate when and where. (Yes, it can get kind of creepy to be recognized by someone you don't remember meeting, or maybe never met. Supers know that too, or they learn it eventually.)

Those abilities are pretty impressive, but they are only the tip of the superpower iceberg. Super recognizers can remember and identify people they've seen before, even many years later. That means that they both remember people over time, and that they recognize people even after they have aged significantly or changed how they look. They are good at recognizing adults whom they've seen as children.[4] They recall almost all the faces that they've ever seen. They extrapolate how people change over time, accurately. That sort of skill is great for reunions and maybe some service professions. But here's where it gets really useful: super recognizers can identify faces even when they are mostly obscured, extremely blurry or out of focus, or a small spot among many. Super recognizers are significantly better than the average population at *both* matching faces from photos and video to line-ups *and* remembering faces from line-ups in video and photo footage.[5] The very best of the best have excellent short- and long-term memory of the faces of others. Which is to say: super recognizers can identify faces (or eyes, or a nose, or ears) captured on camera, even when those faces are doing their best not to be seen.

The History of a Story, Version 1: Richard Russell

As we discuss in chapter 1, the identification of developmental prosopagnosia was the first game changer, indicating that face blindness wasn't just or only the result of damage but could indicate that some capability never quite developed. For those with acquired prosopagnosia, something was broken. It was a pathology. But for those who were born with it, they just lacked the ability. They were worse—sometimes much, much worse—than most people at recognizing faces. Which meant, or could mean, that there were others who were better—sometimes much, much better—than most people at recognizing faces. Richard Russell had a hunch right from the beginning of his time in the Nakayama lab just how much better that could be. But it was just a hunch. His lab already had enough trouble tracking down and traveling to test people who thought they might have developmental prosopagnosia. How could they possibly find time or even find a way to test out his hunch and locate those who had this potential opposite, unnamed set of skills? If even they didn't know they had them, how on earth could Russell?

He couldn't. They found him instead. But then again, he asked for it.

As we've seen, it was hard for Russell and Nakayama and the rest of the team to keep schlepping across the country to test whoever identified as a possible DP. Never mind that not all of them were right; the sheer time, energy, and money just didn't seem to make sense. There had to be a better way, especially since, even at this point, Nakayama figured the number of people affected was likely high, and certainly much more common than they had previously thought, with estimates up to 2 percent of the population. That would mean that one in fifty people were among the worst recognizers of faces. Surely in the greater Boston area there were more than enough test subjects so the team would not have to keep traveling. But how to find them and get them to come to his lab at Harvard? The best way, Russell and Nakayama decided, was to have someone else do the work for them. So they decided to seek out media attention for their work. Russell was careful to emphasize—as most academics are—that the decision to talk to the press was a strategic one. It almost always is: face recognition researcher Josh Davis courts media for awareness and public support, and Richard Russell did it to find local test subjects for prosopagnosia research.[6]

The media gambit worked. From face blindness, super recognition emerged, not just in terms of the spectrum itself, but indeed in how it came to be created and identified. And in their search for prosopagnosiacs the team also found the supers, and, as Russell told me, the two often go hand in hand, living together as roommates or romantic partners and helping one another out.[7] The people who responded to the *Boston Globe's* article on face blindness and found the Vision Lab tended to actually be quite accurate in their self-diagnoses. When they said they would do poorly on the face recognition tests, they were, by and large, correct. Some of them had taken the online diagnostic, and others just knew from a lifetime of struggling to remember the faces of loved ones, friends, casual acquaintances, and in the most extreme cases, themselves. The tests were a highly effective tool to indicate the wildly under- and undiagnosed condition of developmental prosopagnosia. The initial focus of the lab, and Russell's work, was always face blindness. So the initial focus of the strategic media plan was locating people with face blindness. Russell did have a research hunch that the numbers of people affected was high, higher than clinically demonstrated, and he had an equal hunch that there were high numbers on the other end. But he wasn't looking there, at least not yet. So they came looking for him.

To be precise, their roommates did. One of the first identified supers (before they even had the name) was introduced to Russell by his housemate, who was being tested for face blindness. It sounds like a buddy comedy, and indeed it's one of many stories of people on opposite ends of the spectrum coming together. Russell and the test subject were in the lab, and the subject told Russell that his roommate was as good at recognizing faces as he was bad. Russell's first thought was *oh, sure. I'll bet. You just think he's really good because you are really bad.* And that was a reasonable assumption: to people with prosopagnosia, everyone who can recognize faces seems to have a global superpower. You don't have to be off the charts to be impressive at face recognition to someone who can't do it at all. But the subject offered details. Specifics. Skills that weren't just the regular abilities so impressive to those who don't have them, but particular examples of recognizing extras across different tv shows, and recalling faces of casual passers-by, and maybe even knowing someone many years later despite having never seen them in the interim. All these so-called symptoms of super recognition hadn't yet been described. But it was interesting. It was worth a test, Russell thought.

Too bad he didn't have a test to administer. Russell and his lab weren't actually ready for the soon-to-be-named supers. A hunch isn't the same thing as a research protocol. A very strong theory isn't the same thing as proof. And a pair of roommates who are exactly the opposite can't quite be treated the same way. But the prosopagnosia tests were what they had at the time, so the prosopagnosia tests were what they used. And the results were . . . stunning. There were some people who were in fact as good at recognizing faces as face-blind people were bad. But, of course, they were using the prosopagnosia tests, and that could skew results. The materials themselves were simply not hard enough. They were designed to test for people who were really bad at recognizing faces, and certainly much worse than average. What they could say about people who were much better than average was limited. Enough to justify further exploration of what was once a hunch and was now a theory, but not enough to support a definitive claim.

So they made harder tests. And tested the self-identified opposite-of-face-blind people again. (And started thinking about a name, because "opposite-of-face-blind" didn't quite roll off the tongue. Though "super recognizer" doesn't really, either.) There were four people—at least four—who fell into this yet-unnamed bucket of people with extraordinary face recognition abilities, leading to the landmark 2009 paper that launched a small but vibrant super recognition industry. It's actually a rather modest essay for something that goes on to have such interesting practical and theoretical consequences. But they wrote it quickly, in a bit of a frenzy to put their stamp on the discovery and keep the research going. Though, in the case of Russell, not by him.

But before they could publish the paper, they needed a name. They needed a good name, or at least a good enough name. Russell doesn't think he's very good at naming generally, and this case presented a particular set of challenges. It had to tie into face blindness in some way. It had to be descriptive enough that people knew what this new thing was. It had to be available. (For this last, everyone with a start-up in a garage or a new product can sympathize.) "Have you ever named anything?" Russell asked me, a bit plaintively. "Only my kids," I answered, "and that was much harder than I expected, mostly because my husband kept coming up with ways that a name could be mocked." (I didn't learn a lot about naming, but I did learn that British schoolchildren have a hitherto unsuspected capacity for great cruelty.)

I like my kids' names quite a lot; Russell feels more ambivalent about what he came up with for his research baby. "It's not bad, it could be worse," he half laughed. It does what it is supposed to do. But he never felt like it was (and I could hear the capitalization in his voice) IT. He had to compromise: at first he wanted a Greek term to resonate with prosopagnosia, which makes good sense both in terms of the origin story and in terms of research protocols. A Greek name would resonate with the medical model of face recognition and tie it directly to face blindness. It's also more universal and would work across languages (a problem that Harry Potter translators certainly encountered multiple times when—spoiler alert—trying to scramble from "Lord Voldemort" to "Tom Marvolo Riddle" in a way that made sense). He spoke to friends, trying out options. He reached out to Greek scholars for suggestions. Nothing quite worked.[8]

In the end, Russell settled on super recognizers as a reference to super tasters. I was surprised; the two sets of super skills are dramatically different, and honestly, being a super taster sounds pretty terrible. Sweet stuff is too sweet, and vegetables are too bitter. Spicy food is impossible. It's biological, and straightforwardly structural: super tasters literally have extra taste buds. That's why they taste more.[9] Super recognizers don't have extra rods or cones or really extra anything in their eyes, and maybe not even in their brains, though it is likely there is some neurobiological factor that has not yet been identified. Supers do show greater bursts of electrical energy when processing faces, and some research indicates that it might be heritable and thus genetic.[10] Russell liked that super recognizers is straightforward and accessible to English language audiences. The name actually gives a sense of what the condition is, unlike prosopagnosia, for example. It's down to earth. Maybe a little prosaic. Maybe not quite specific enough to faces. But overall, it "feels like it's okay."

Honestly, it wasn't quite the naming story I was expecting. And another shock: Russell didn't even remotely intend an association with superheroes. And he didn't even remotely foresee that others would make that connection. The truth is, he told me, "it's a mystery why people are so interested in superheroes."[11] And he meant the comic book and movie franchise kind, not the super recognizing kind (though he's less interested in those these days also.) He genuinely didn't think that super recognizers would be thought of as superheroes. He didn't think their skills would be applied to, well, anything in particular, and certainly not to law enforcement. When it came to the superhero implications of super naming, he really didn't think about it at all. But of course, that doesn't mean that others wouldn't. And of course, others certainly did. And do.

The initial four supers provided great early evidence for the existence of these abilities, and by implication, great early evidence that to date, face recognition theory had more or less got it all wrong. When there is something that has extremes on either side that include smaller numbers of folks and then a great big arena in the middle that captures most people, we call it a spectrum. Though it's not a perfectly balanced one, as we've discussed. The statistic that gets thrown around a lot, even by me, that the top 2 percent of the population are as good at recognizing faces as the bottom 2 percent are bad isn't quite accurate. Yes, it's a spectrum, with a very few people at the extremes and a big heavy middle. But there are a lot more ways to be bad at recognizing faces, and a lot more reasons for people being bad at it, than there are ways of being good at it. For example, many people on the autism spectrum have trouble recognizing faces. They tend to score pretty poorly on the tests that identify prosopagnosia through low results. But that kind of trouble with face recognition is likely not the neurological opposite of super recognition. It's something else that produces similar results. So, in Russell's terms, the lower tail on the spectrum graph is almost certainly fatter than the upper tail.[12] But it was still a spectrum, nonetheless. Not, as neurologists and psychiatrists and other face blindness researchers had thought to date, a pathology. And that changed everything.

Think about it. This isn't just a question of technicalities, or even a way to challenge the traditional reliance of neuroscience on brain scans and lab tests and large-scale scientific studies. When we think of something as a spectrum instead of a set of binaries in which one end is a disability, a pathology, a sign of brokenness, we are insisting that even for those who are not the extreme ends of a given facet of human experience, we do not all see the world the same way. Spectrums are sticky entities: while in trans studies and sexuality studies more broadly, the spectrum is deployed in service of fluidity and flexibility, it often works to reinforce static categories and the distance between poles.[13] The face recognition spectrum places blindness and super-ness in opposition, and there is a loss there, in both how we understand how we relate to others and what people on extreme ends might share. We are not attentive to extremeness itself as a shared entity.

The emergence of super recognition as a category is also a creation, and it is also something that gave name and shape to something that was always there and always hard to see. Naming and framing hold great resonance: superpower rhetoric stands in sharp contrast with prosopagnosia and its status as a disability. For many, face blindness isn't all bad, and super recognition isn't all super, and perhaps, despite the obvious differences between the two, there are things they have in common beyond the difficulty in identifying them. The two extreme ends of the face recognition spectrum certainly have vastly different social, interpersonal, and emotional implications for those on either end and for those around them. Let me be clear: people with prosopagnosia face serious challenges that differ significantly in scope from what supers have to contend with. However, despite the fundamental differences, people on both extremes share a lack of relationship between recognizing someone's face and having a relationship with them.

Spectrums by definition entail extremes, and it is those extremes that we often focus on. We get locked into the poles and forget about the fluidity in between. The poles become the outliers—which of course they are in a way—and we can easily imagine that what lies between is all the same. This approach erases the diversity of experience across and within the extremes. It's hard to understand that people's experiences are different than our own. And with something as intuitive and ordinary and undescribed and tacit as recognizing faces, it's nearly impossible. People with average or above average face recognition abilities truly cannot understand what it is like to be on the face-blind side. There are descriptions about it. But I at least struggle to understand them, because I generally look at faces and recognize them, and I don't quite know what it would be like for that not to be the case. And more than that, until I started thinking about it, I never thought about it.

As Russell said to me, we don't even have the words to discuss how we perceive and recognize faces. We lack the language. We lack the vocabulary. We lack the experience. But now, at least, we have a framework. Now at least we know, even if we can't understand, that when people look at one another, we all see and perceive and recognize differently, and to a different extent. Which, again in Russell's words, "leads to broader possibilities of seeing many different worlds when we are looking at this world."[14] And that's what drives Russell's research now: thinking about the often subtle and undetected differences that go into how people perceive others and the world around them, something he's passionate about. So passionate that he walked away from what turned out to be fame, (modest) fortune, and a lifetime supply of research subjects. Because that 2009 paper was just the beginning, even if, for Russell, it was also the beginning of the end. But he had to see it through, at least to the point of establishing a framework and a research agenda. And then, only then, could he walk away.

He needed more subjects to see it through. And many more people with these aptitudes would go on to be discovered, though not quite as many as think they have it. "It's funny, kind of interesting" Russell mused, "though maybe it makes sense in hindsight. When people self-report being face blind, they would basically always perform badly on the tests. They are pretty accurate judges. But," he continued, "we never got a super recognizer just from talking to them personally." Unlike people with prosopagnosia, "people who think they are really good think they are better than they are." And that could be because being really good is actually quite impressive, even if you aren't in the super category. Russell suggested another reason: "Americans may be more willing to be good at something than bad at something," so they are always willing to assume they are better than they are, but rarely make themselves out to be worse than they are. If someone (in the United States) claims to be really bad at recognizing faces, believe her. She's probably right. But if she thinks she's really good at it . . . she might not be. (I immediately asked if people in the United Kingdom were better at self-diagnosing super skills. Russell didn't know.)

Building a Super Recognizer Empire: Josh Davis

Dr. Josh Davis might have a better idea if the supposedly more modest Brits are better at knowing when they are actually good at something. He's British, after all. And he spends *a lot* of time with super recognizers as part of his research at the University of Greenwich. For those projects, he's interested in anyone who does especially well on the online tests he and other have developed, including the CFMT (Cambridge Face Memory Test) and the GFMT (Glasgow Face Matching Test), on which it is possible for a very few people to achieve a perfect score.[15] A more recent addition to the super-testing arena, the UNSWFT (the University of New South Wales Face Test—quite an unwieldly acronym) is even more challenging, and to date no one has gotten 100 percent. Researchers Dr. James Dunn, Scientia Fellow Dr. David White, and Dr. Alice Towler launched their test in 2017 to provide even more precision in ranking super skills, using it in combination with the CFMT and the GFMT to shift the scale; in their system, those scoring over 70 percent are super recognizers, with ample room at the higher end for the superset of supers.[16] All these tests are based on principles that are similar to the prosopagnosia screening mechanisms we've already encountered. It stands to reason: they are measuring the same thing. It is simply a difference of quite considerable degree along a spectrum of the same kind. Davis's online page has garnered more than 7 million respondents to date, and he has a huge backlog of research results that he has yet to publish.[17] A great number of experiments and analyses are currently in process in his lab, one of the leading face recognition labs in the world, with a particular expertise on the super end of the spectrum. His group publishes regularly, alongside fellow British super recognition researchers Sarah Bate at Bournemouth University, Ashok Jansari at Goldsmiths, and many others around the world. There is global interest in the supers and their surveillance potential: facial recognition is big business. But it's no surprise that so much of the early research was clustered in the United Kingdom, a country that bet big on cameras and seemed, for a long time, to be losing. Until the face recognition technology (FRT) got much better, the supers seemed to be the way to make the CCTV pay off, and there was a lot of government and private investment in finding out if that was the case.

In fact, Davis's face recognition work has brought him a robust research profile and some media fame and (modest) fortune. That's in small part through his academic post, but, as he has written, despite the clear application to his work, his lab remains underfunded and without significant grant support.[18] His more traditional financial success comes from his other job—the one where he needs not just the top 1–2 percent of face recognizers, but the top 0.1–0.2 percent. It's actually not as hard as you might think to recruit people and find super recognizers. He needs those super recognizers for his academic research, for his peer-reviewed papers and REF (Research Excellence Framework) outcomes (the British system for evaluating university research outputs). But that top 1–2 percent isn't quite good enough for his other pursuits, or his other partners, which include police forces, unnamed international government agencies, security firms, software design companies, and almost certainly a number of spies. Josh Davis is helping to build a super recognizer empire.

Davis has partnered with Scotland Yard to identify and train the elite squad of super recognizers to make identifications of criminals based on visual footage. He started looking into supers in 2010, based on some planning he'd done a few years prior, when he didn't even know what to call the group of people he wanted to study. His research was casual at first: he told me that he "texted a few people who are now called super recognizers," but he didn't really have a lot of funding and found it hard to find someone to give him the money to do what he wanted. The funding problem changed dramatically when he connected with the Metropolitan Police (the Met) in 2014. Davis gave the Met empirical support to their claims. The Met gave Davis a media story, a practical application for his work, and, eventually, a substantial amount of money.

In the intervening years, Davis has leveraged the huge success of that endeavor to help build Super Recognisers International with Kenny Long and Mick Neville, two key players in the original super squad at the Met. Super Recognisers is essentially an agency that identifies super recognizers (yes, through those online tests and the subsequent follow-ups for those who score high enough and want to spend hours of their lives taking them) and then places them with various companies who have need of their skills.[19] The gamut is still pretty narrow: we haven't, to the frustration of at least some supers, figured out all the ways to monetize and employ these skills.[20] But security agencies and overall event staff clearly benefit from employing people who can recognize repeat criminals at a glance.

Davis is also experimenting with expanding the range of possibilities for super recognizers (and for his database of super recognizers). His latest venture, called Yoti, offers a system of visual identity verification so that you can use your selfie to prove who you are. There are endless applications: alcohol and other age-sensitive purchases are obvious, but also financial transactions where identity is paramount. And there are social benefits as well: Davis explained how Yoti's technology can be used to keep adults off children's social media sites. Overall, it's about gatekeeping: letting people in and keeping people out—using pictures of their faces.[21] As we'll see in chapters 5 and 7, the history of face recognition has always been intertwined with the history of surveillance and the history of biometrics; computing didn't invent the process. It is, however, continually refining it. And while Yoti, like a number of other organizations, contains a section on their website about ethical principles and tech for good—and they are genuine in their support— the larger framework of racial capitalism always limits the extent to which any of these principles can be applied, or whether they can be applied at all.

Yoti gives organizations access to the information they need to verify someone's identity. Although Yoti is far from the only company doing this, its special ingredient is super recognizers who scrutinize the original images provided by people for identity verification, making sure that they are valid and making sure that the algorithms in the computer recognition systems are working correctly. Like any start-up, if it takes off, the rewards could be significant. It's a huge international project, and so Davis needs supers from around the world who can work with all kinds of faces. And it means he needs the best of the best. Most super recognition tests, he explained to me, test short-term face memory. That's the standard. But Davis needs people who are excellent at both long- and short-term face memory. He needs people who are excellent at spotting faces in a crowd, and he needs people who are excellent at face matching. He needs people who are excellent at every aspect of face recognition. Even most supers fall short on some aspect of the work, and most labs investigating super recognition are somewhat limited in their testing abilities. And of course, that's true of all face recognition software: we always need people at its core.[22]

Much like Russell did in 2009, super recognizer researchers have to keep making new tests to make sure that they are hard enough to differentiate between the merely amazing and the superlative, and that they test real-world experiences and applications.[23] Josh Davis in particular has had to make tests that may examine similar things but are customized for various applications. Given his international focus and partnerships with governments around the world, Davis needs tests that explore recognition among different races and ethnicities. The online test that most people take in the first instance features only white faces, but super recognition of white faces isn't that useful in some instances. Davis has developed a series of tests using different ethnicities, some of which test abilities across ethnicities as well. For his work with a government agency in Singapore, for instance, Davis developed a test that took into account the three major ethnicities of the country—a critically helpful customization from a research perspective. Davis was also able to answer a question I keep asking about the other race effect (ORE, the phenomenon whereby people are much better at distinguishing faces of those from their own race or the one they are most used to seeing) and super recognition. Davis said that, yes, supers are generally as susceptible to the ORE as other people, though real-life results from police working in diverse areas show that many of the supers on the squad are still excellent at applying their skills across races.[24] He did the work because he's interested in it, but also because he needed to know the results for his international partnerships. It is in fact the case that the one fuels the other. The one also funds the other.[25]

Davis's story effectively illuminates the challenges of doing certain kinds of expensive research in the current academic environment, particularly in the cash-strapped British context where every single piece of academic work is evaluated based on specific metrics that require hard numbers.[26] Davis has figured out how to make money from his life's work and passion projects, and that's great as long as he is comfortable with how it all plays out. While he is bound by certain nondisclosure agreements and confidentiality requirements, his collaborations do not require him to change his data or manipulate his research. If there is material he can't talk about, that's not terribly uncommon. And, in his case, the problem isn't so much what he can't talk about; rather, it's finding time to share what he can talk about.

Davis's plan was never to be on television shows and give media interviews and collaborate with business tycoons. But he doesn't really mind; as he told me, he has a background in acting and music. He didn't need media training, and he is comfortable in front of the camera. To be honest, I guessed that might be the case both from watching his appearances and even chatting with him on Skype. A friendly, charismatic man in his mid-forties, Davis immediately put me at ease and made me feel like I was doing him the favor by having this conversation. I can see how he moves easily between the multiple worlds he inhabits: lab boss, researcher, collaborator, writer, talking head, interviewer, businessman, job recruitment specialist. He's approachable and affable.

Like Russell, Davis wants me to understand just how committed he is to the work. And like Russell, he's had to make some compromises, but of a rather different sort. When I ask him about the television appearances, he's at pains to make me understand that what we see is not what he gave necessarily. The media (of course) edit heavily to make a story, so his knowledge and his interests are not what are emphasized. In his words, "the media cut around the bits and pieces I want to get across, and put across what they want to get across." But that's part of the price he pays. And he knows what it looks like from the other side.

Despite now literally working for and with the police, Davis started his career working against miscarriages of justice, researching the unreliability of witness identification and lineups. His research was always about how people recognize faces; in the criminal justice prosecution context, the answer is often "not very well." He wanted to find ways to make that better. Instead, he found people who can do it better. And he hasn't given up on the criminal justice research, he insists. I asked, thinking of Russell: "Do you feel stuck now with the super recognizers?" He's not. He has his growing super empire, but he also has two PhD students working on the possibility of super voice recognition, and other projects on consent, rape, and prostitution. He's got projects on the psychology of the death penalty and gun ownership, and he remains deeply committed to finding ways to minimize the miscarriage of justice. He's also actively working with governments to participate in his ever-expanding project: to date he's working with Australia, New Zealand, Germany, and Singapore. He supervises graduate students working on voice recognition to supplement the face recognition infrastructures, and, possibly, to provide alternatives for those who are leery of being tracked visually but may somehow be more comfortable with audio surveillance. He keeps trying to recruit US agencies and police forces and has spoken to them at their invitation, but to date "I failed to convince them." Davis suspects that American organizations are happier to use facial recognition technology over people, thinking that "if a computer can do this job, why do I need a human?"[27] I wonder if there isn't more to the reticence of US agencies to enroll super recognizers into their policing and surveillance; while Americans have been happy to embrace cameras for the purpose of private enterprise, they remain suspicious of many forms of facial recognition in criminal justice contexts. As we'll see in chapter 4, there has been a lot of resistance to face recognition technology; perhaps there is a concern that rather than alleviating those pressures, super recognizers might make it worse. Superheroes may be far scarier than super computers.

Tempering the Hype: Jennifer Jarett

Jennifer Jarett knows a lot about navigating a nameless weird condition that no one else seems to have or understand. Jarett is friendly, warm, and extremely engaging. She has done scores of interviews after the publication of the 2009 Russell, Duchaine, and Nakayma super recognizer paper. I got the sense that following that initial rush of attention, Jarett sort of stopped thinking about being a super. It was a feature of her identity, perhaps like being from Philadelphia or having copper hair, but it wasn't particularly central. When I asked her if her super recognition abilities came up on dates, she smiled ruefully and said, "I tend to think like if it's coming up on a first meeting with someone, then it means it's not going very well, because we have nothing to talk about."[28] It's a kind of party trick that she can trot out, but it's not, in a meaningful sense, who she is.

Her super recognition ability has, however, framed how she interacts with other people. Before she had a category, a label, or even a model for how to think about recognition, Jarett was good at this. Eerily, spookily good. She "always knew who everybody was," both in real life and on screen, and she was able to say "that person was in this and that person was in this," even across years. Her family would ask her, "Do you know who that person is?" and she'd say, "That person was in a commercial however many years ago." Sometimes it made for a funny story, like when she recognized someone on a flight when she was 16 as a mildly famous person—in her words, "the kind of person who would be in the first two minutes of *Murder, She Wrote* and falsely accused of the crime." Her family didn't believe her, but she insisted on getting his autograph. That fall, they were all watching the season premiere of *Family Ties*, an episode focusing on book burning, and sure enough, this fellow passenger played the principal character. And Jarett was triumphant, telling her family, "You can never doubt me again." As we'll see in chapter 4, recognizing extras and minor characters is a consistent feature of super recognition, and one that was all the more notable in the days before IMDb (the internet movie database).

That story gave Jarett bragging rights, but "it was still not quite like I have something going on here." It only really "hit home when I started college." That's when things got . . . awkward. As a first-year undergraduate, Jarett was "meeting people all the time," since "everybody introduces you to everyone around them." Most of us mostly remember those we meet in that kind of situation; Jarett "would remember the most minor interactions . . . but other people wouldn't." She "started complaining to my friends that people are so fake." She gave a particular example that has stayed with her all this time, remembering when she was "at a party and someone said, do you two know each other?" Jarett, of course, "was like yeah I remember the first day of classes, I came into the cafeteria and you were sitting at a table with all these people." Jarett got even more specific, recalling that "everybody introduced us, and then you said you have to go study for a bio exam." While others might remember the incident as one among many, it would be less common to connect it to each of the faces around the table and then recall the two pieces together later. To Jarett, recalling faces cements the moment, such that "these were all memorable encounters," and she kept feeling hurt and excluded that others did not have the same recall, until "my friends were like there's something wrong with you." They emphasized that "it's not them it's you," and in fact they "would not remember either." They were puzzled and did not "know why you think people are pretending not to remember you" as "these are not memorable experiences."

But to Jarett, of course, they *were* memorable experiences. They were exactly as memorable as those experiences that people did remember, and she could draw no distinction at all between them. It was only when "I got to know these people who I thought were being fake" and they "became friends of mine" that Jarett "started to really know them and realize they weren't pretending." It took until "the second semester of college" for her to adjust her behavior; she learned to "be at a party and someone would say 'Do you two know each other?'" If Jarett had met that person even fleetingly, she would want to say "yes," but instead, "I started to wait for the other person to answer."[29]

One of the pitfalls of being a much-profiled super recognizer is that such anecdotes can become rather well known. She shared the bio test story with *60 Minutes* and felt so guilty that she reached out to the person in question to assure her that "you're so sweet and so kind" and "of course you weren't being fake at all" because "you're the least fake person on the planet." In fact, it was precisely that person's sweetness that "helped me learn that there was something weird going on here." Jarett points to that moment as "the sea change for me in terms of realizing there was something going on with me that was not going on with other people." It was not quite the lived experience of a lifetime that convinced her, but the combination of a particular person, the feedback from friends, and the sharp contrast between expectation and experience that caused her to challenge her own conclusions. It sounded rather destabilizing, actually, and it put the moment of explanation in sharp (and palpable) relief.

The relief didn't come for quite some time. But it happened almost exactly the way Russell and the Vision Lab team hoped it would. Sometime in "the mid 2000s," Jarett was reading "the *Sunday Times Magazine*" and "they talked about prosopagnosia," which was "sort of new at the time." The article, which was part of the "Year in Ideas" feature "explained what it was, and they talked about a study that was going on at Harvard with a couple of people," including Richard Russell. They couldn't have scripted it better: Jarett "emailed him and I said you know I just read this "Year in Ideas" thing about prosopagnosia." Jarett had a specific question, telling Russell that "I'm wondering if there's an opposite of that because I think I have it, if it exists." Unsurprisingly, Russell "wrote back quickly" saying "we don't know if there's an opposite, but we think there might be." The team had to develop a way to assess that, so they were "putting together a test" and wanted to know if Jarett "was interested."

She was. In fact, she thought *oh my God, that sounds so great*. Jarett's sister thought that she herself might have the super recognition skill too (though her mother, "to her credit when I told her," said, "It's you, it's not your sister"), so Russell decided to test them both. And despite the media strategy designed to limit travel, he and Jarett coordinated "a time when I'm home visiting so you can test me and my sister at the same time." It was not as inconvenient as it might have been, as "Richard's in-laws live in Pennsylvania, so it was sort of nice." Jarett did feel a bit bad about the travel, especially when her sister's test came back showing average recognition. She started to panic that "you've come here to test me and what if I end up being like my sister." Russell reassured her that "you know with the research that we're doing, we need to have a ton of people taking this test." He emphasized that "it's worthwhile anyway, we learn something about every person who we test so don't worry about it." That was comforting. More comforting still was when "he did the first test and it ended, he patted me on the back and he said, 'You were worth waiting for.'" This was a huge moment for Jennifer Jarett. For her, "it was so exciting." She "really just thought, thank God I'm not crazy." Until that moment, she really didn't know for sure. It "wasn't written about anywhere," and she had begun to wonder if she was "imagining this." When Russell told her that "you were worth coming [to Pennsylvania] for," she "did not think it could get any better than that." She was just "so thrilled to have a name to put to it" and to know that "this is a real thing."[30]

Jarett's version of the story from here echoes Russell's narrative; he reached out to her to do various media, and she agreed to go on BBC radio, followed by *60 Minutes* and an interview with the *New York Times*.[31] And then Jarett got picky. And maybe even a little bored. To her, it was "not that interesting." As she put it, "It's what I've lived with and so it's sort of like bragging to a colorblind person about being able to see color." It mattered to her to help publicize it so that people would know it existed, but at a certain point the specifics just weren't all that relevant. So when, say, the game show *Superhuman* asked her to participate to show off her skills, she just wasn't interested.

I pushed her on this a little. I'd seen the episode of *Superhuman* that features a super recognizer.[32] It's pretty amazing: the contestant looks at an entire audience. He turns away. Everyone moves seats. Someone leaves. He identifies who it was who was no longer there. I admit I was impressed. Jarett was not. "It's fake," she declared. She knows because "they reached out to me to do it and I said no." It was too nerve-racking to be tested in that way, as she knew from experiences on other shows. But they kept coming back to her and offering to make the audience ever smaller in order to rig it. She kept saying no. *60 Minutes* had asked her to do some on-the-spot identifications and she did "not want to take a guess"—because she wasn't positive.[33]

While the entire interview was fascinating, this was *really* fascinating. "You didn't know for sure?" I asked. And Jarett was very clear to say that "it's not 100 percent." In fact, "it's sort of apocryphal to say" that super recognizers have "a photographic memory for faces." Jarett's really good. In the original paper that Russell published, she was the best by far. In the subsequent years and ever more refined tests, super recognizer researchers have found others who are also extremely good at recognizing faces. But, at least according to Jarett, almost no one could do what these quiz shows and maybe even these police recruiters and security forces claim. It just doesn't work like that.

Jarett is skeptical of the claims of the police forces and security teams and governments. "I need to see a video of someone at the very least," and it needs to be "substantial video, so if it's grainy from miles away" like a lot of "security footage," then "there's no way I could ever" identify someone. She specified that "I would never in a million years feel comfortable doing it even once" for the purposes of criminal prosecution. To Jarett, super recognition isn't really a superpower. She's not being modest or downplaying her abilities. And it's not that she's not a super recognizer. She is. She just thinks, as she said, that "it's not that interesting."[34]

Part of the problem, to her, is the name, which is "pretty goofy." She'd rather have had it be "the converse of prosopagnosia," which is probably a more accurate, if less resonant, way to frame these abilities. She acknowledged that "it certainly has captured people's imagination," but she's not really convinced that she has a superpower beyond being the reason that the notion of a face recognition spectrum was solidified. Which, in and of itself, is a pretty big deal. "They used to think it was like everybody saw things the same except for people who couldn't remember faces." But after Jarett and others, "they realize no, it's a whole spectrum." And because of that, when Jarett was first identified she told her friends that "they haven't found a ton of people like me yet, but they're going to."

And they did. But they are still looking, and still learning. They had to refine their tests to reflect the populations in which they were recruiting: if the tests were all white men (as they originally were), people in non-white majority countries were going to do worse.[35] They had to adjust for cultural bias: the "before they were famous" test shows pictures of celebrities as children and asks people to identify who they are. If you don't know them as adults, even the most super of recognizers won't be able to identify them as children. But even as researchers find more and more super recognizers, Jennifer Jarett's caution holds strong. She warns us not to be distracted by the name super; for her, what is most super about it is the name itself. The rest, she insists, is just not that interesting. Certainly not interesting enough to mention on a good date, let alone to compete in a fake game show, or—and with greater stakes—to identify someone as having committed a crime, or, in a poignant example that she shared, help a Holocaust survivor identify a child who may or may not have been rescued based on one photo. As Jarett wrote to that survivor, "I'm so sorry I cannot help you." But she couldn't. It "really just broke my heart but you know, I really can't tell from one photo what someone looks like. I really just can't."

And if she can't, probably no one can.

Conclusion: Super?

Richard Russell is one researcher who worked with a team of other researchers and identified four specific people as part of a larger scholarly agenda to study a condition that, in the context of these networks, he and others framed and named. Through his research and with his team and subjects, Russell invented super recognition. That doesn't mean people like Jennifer Jarett weren't already doing it and people like Josh Davis weren't already studying it: clearly, they were. It just means that they didn't have a name for it, and without a name, it was very hard to categorize it or even know it was a syndrome. As we'll see in chapter 4, there is a lot that connects super recognizers both to one another, and, in perhaps surprising ways, to face-blind people. In this chapter we explored a local set of networks and people, and, in so doing, helped trace the global story. Next, we'll see how it worked on the ground. And while we will continue to think about how super recognition and face recognition more broadly are studied, we will also be attentive to what is not studied, and why. We'll dive into the notion of spectrum thinking about how poles become sticky, with blindness at one end and superness at the other. And we'll stop to ask: What is the opposite of super, and what are the stakes for that opposition? And also: Is being a super recognizer so very super?

A Super Useless Super Skill

Meet the Supers

What happens when you invent a super category and people discover they fall into it? For most supers, not much, as it happens. Recognizing people on the bus or Brad Pitt's baby pictures is a cool party trick and learning *not* to always recognize people on the bus or strangers' baby pictures is an important life hack, but (unlike prosopagnosia), naming doesn't seem to change much for those who get it. As Jennifer Jarett put it in chapter 3, it's cool, but far from the most interesting thing about her. Unlike the understanding or sympathy that might be aroused by the identification of one's face blindness, being categorized as a super recognizer doesn't come bearing the same sense of relief. Supers don't really need accommodation or understanding beyond the creepiness that they soon learn to hide. The skills don't seem terribly valuable, and might even have some downsides, not that anyone's really looking for them. What face recognition researchers, media companies, governmental agencies, and indeed the supers themselves are busy paying attention to is, of course, surveillance. Because while recognizing extras on television shows is fun, identifying perpetrators caught on camera is actionable. For a long time, this fact was particularly relevant in London, with its hundreds and thousands of cameras everywhere. But with the growth of face recognition technology, masters of surveillance might be needed everywhere. At least, Super Recognisers International is hoping that's the case. They may be wrong. But they did get a good head start by being based in the United Kingdom.

A *New Yorker* article from 2016 chronicled how Scotland Yard uses supers to track down hundreds of previously unidentified criminals who were captured on London's ubiquitous CCTV.[1] As anyone who has watched modern

British crime television knows, Britain is possibly the most camera-monitored country in the world.[2] London alone was blanketed with over four hundred thousand security cameras, containing 50 percent of the world's closed-circuit television as of 2017.[3] Britain's massive surveillance system trades privacy for the promise of increased safety through screen monitoring at tremendous financial cost.[4] These recordings are always being evoked and called up on said television procedurals, producing images that often catch the criminal in the act, or, in more interesting plot twists, introduce another villain or another crime entirely. But in television as in life, the recordings can go only so far: seeing an already-reported crime being committed on screen only confirms that it happened. It doesn't actually produce the criminals, merely a recording of them. That recording then has to be correlated with an actual and identifiable person whose photo is on file or who is known in person in order for the suspect to be apprehended, or even simply questioned.[5] Face recognition software can match two photos taken in ideal conditions, including good lighting and full-frontal facial views (like a mugshot) but has trouble correlating a single visible feature (like an eye or an ear) that is captured on video with a different photo. It has often misidentified gender, basic characteristics, and even species, particularly among people of color, as Joy Buolamwini's work with the Algorithmic Justice League has demonstrated, and as we explore in chapter 6.[6]

There are flaws in the system, though the recent (and privacy-violating) harnessing of literally billions of photographs of faces has made the face recognition technology (FRT) much more encompassing and robust. Companies like Clearview AI scrape all the images online to which they are legally entitled (a number approaching 20 billion), and that activity has changed the game for face recognition software.[7] While other tech giants including Microsoft, Google, IBM, and—very recently (and only sort of)—Facebook have opted out of the face recognition game, our images are basically all part of the database now. So FRT works much better than it used to. But it's still not perfect. Even as the tech has continually refined to produce 3-D images from static photos that dramatically increase the data-matching capabilities, individual features remain a challenge for the technology, which for a long time wasn't very good at invading people's privacy.

Enter the supers.

The formation of the "super recogniser squad" in Scotland Yard in 2012 generated significant media attention around super recognition. That attention, along with easily accessible diagnostic tests online, caused many people to discover that their uncanny knack for never forgetting a face actually has a name, and that the name is actually a diagnosis.[8] And that diagnosis places supers at the extreme end of the face recognition spectrum. Like all spectrums, there is a continuum with another extreme: in this case, face blindness, or prosopagnosia. While these two extremes present as complete opposites, we have much to learn from thinking of them together about de-centering the primacy of face recognition as a measure of emotional connection between two people. This practice challenges the binary opposition between the far reaches of a diagnostic spectrum and opens up space to explore what may link these extremes. In this case, in addition to common challenges in naming and identifying recognition-related outliers, we find similar relationships between memory and recognition at either end—namely, none. We also see social tendencies that in both cases seem to be motivated by recognition: either its lack or its completeness.

We have already met Richard Russell, Josh Davis, and Jennifer Jarett, who taught us about the process by which super recognition emerged as a category and as an experience. These firsthand accounts situate super recognition in time and space; here, I turn to the larger infrastructure around super recognition, exploring its application in policing and surveillance. I track people who learn about super recognition as a personally relevant category that may—or may not—change much for them. I examine published papers on the topic in the scientific literature; there are no studies to date on the psychosocial implications of super recognition aside from evaluating if they have identified their ability.[9] As I turn to popular representations of the ability in order to understand how it is discussed and rhetorically framed, I search for every mention of super recognition in news media and television. News articles fall into three broad categories: they report on the discovery of the condition in 2009; they discuss the use of super recognizers in Scotland Yard from 2012; and they tout the availability of online diagnostic tests so that anyone can see where they lie on the face recognition spectrum. In all the articles, super recognition is presented as a kind of superpower or super-human skill; this presentation is echoed on television shows, which feature super recognizers doing astounding feats of identification. I also examine the television features on super recognition and analyze their presentation of and rhetoric around the skills and how they can be applied. These shows highlight the consolidation of super-ness and criminal detection specifically, and doing *something* applied with this skill more broadly, further underscoring the binary between face blindness and super recognition.

The Stickiness of Spectrums: Or Thinking across Extremes

At both ends, face recognition is the same kind of index to the relationship of the recognizer to the recognized: it provides little to no information about the emotional connection between the two. That doesn't mean that they (like all of us) *feel* the same way about strangers and intimates, but that they recognize them equally well or equally poorly. The most face-blind people can recognize no one; mothers, lovers, and acquaintances are all equally visual strangers, regardless of memories, shared experiences, and past encounters. Super recognizers recognize everyone equally; mothers, lovers, friends, and strangers all register visually exactly the same, which is to say, entirely. For super recognizer Duncan, his social skills have been a professional boon in his field of recruitment; the creepy factor is minimized in this context and people are instead "extremely flattered when you remember them, or if we haven't actually met in person and I've only seen their picture on social media."[10] A prosopagnosiac may retreat from public social encounters in order to avoid causing offense or feeling stupid. Not recognizing others can present a serious barrier to building relationships, though the online world has certainly opened up new avenues of encounter, often with names and identities attached. Equally, as Heather Sellers has noted and we see epitomized in former governor and current senator from Colorado John Hickenlooper, face-blind people may become experts in creating instant intimacy, connecting to each person as though they have a long-standing relationship, because they just might.[11] Memories of others, at these extremes, are meaningless. As an index to the depth and meaning of the relationship, for both prosopagnosiacs and supers, the face and its recognition are blank.

The two sides of the face recognition spectrum share much: the meaninglessness of face recognition as an index to depth of relationship; the difficulty of identifying and naming the condition; the need to self-consciously learn social cues to compensate for their recognition abilities. On both extremes, there is what I call diagnosis without disease, creating subjects rather than patients: no treatment, no medical insurance implications, no real impetus for patient mobilization and advocacy beyond raising awareness and providing support.[12] This isn't, yet, a social movement: not until there is something clear to be gained.[13] The researchers benefit, of course; they need these subjects to further explore something that is hardly understood to date. The collaboration remains largely unidirectional, with researchers studying, framing, and describing subjects; not yet are these not-yet-patients coproducers of the conditions. Not yet in the process are the patients code-termining the categories and working collaboratively with the researchers around their conditions.[14]

Of course, the stakes are much higher and the challenges much greater for the face blind, but that doesn't mean we should ignore the psychosocial consequences of being a super. We do ignore them, as I show, partly because of the culture of diagnosis and disability: when something is understood to be a lack, we look closely at its disadvantages. When something is a super-power, we think much more about how to monetize and apply it. We don't always stop to ask: Is it so super? That's the trap of what I call spectrum think-ing: while there is great space within a spectrum for flexibility, the extremes remain binary poles. And if one thing is bad, the other must be good: as Anna McGuire has argued, the extreme end of the autism spectrum mobilizes nega-tive and even violent associations that attach to a range of behaviors, with the notable exception of savants, who can be thought of as "supercrips" with ex-traordinary abilities.[15] But supers do not fall into the offensively inspirational trap that scholars have identified around supercrips, highly athletic disabled people who are held up as impossible standards *or* disabled people who are presented as admirable simply for existing, because they are not, by defini-tion, disabled. When it comes to that which is framed as a disability, a defi-cit, an absence, popular attention is quick to point out what is bad, and how it deviates from (and contributes to) notions of normalcy.[16] Spectrum extremes are rigid, and their labels are sticky.

And oppositional.

It took sixty-two years after prosopagnosia was named for super recog-nition to be identified clinically in a peer-reviewed publication, in part for similar reasons: much like prosopagnosiacs, super recognizers just didn't re-alize they were particularly unusual, so they did not connect with research-ers and doctors around their abilities. They knew they were rather good at something, and better than most people around them, but they didn't actu-ally know that it was a *thing*. Neither did the researchers who worked in face recognition, who thought that face blindness was a syndrome or lack rather than the extreme end of a spectrum. It wasn't until 2009 that Richard Rus-sell, Brad Duchaine, and Ken Nakayama named the first four clinically acknowledged supers, who in fact self-identified. These four presented themselves for testing after hearing about the lab's prosopagnosia research and claiming they manifested the opposite condition.[17] And, as we learned in chapter 3, they were right.

With the self-identification of supers, face recognition was reconceptualized as a spectrum; a major neurological approach was overturned through people knowing themselves and their abilities rather than brain-scanning or lab tests, though many scholars speculated that these abilities existed.[18] Until that point, face blindness was categorized as a syndrome or pathology that a very small number of people had. With the reconfiguration of face recognition as a spectrum, researchers reframed their understanding of what face blindness is and how many people have it. That's how spectrum thinking works: with the entire population now situated on the face recognition spectrum, everyone falls somewhere, though in this case the bottom tail is fatter than the upper. There are more ways to be bad at recognizing faces than to be good at it, so there are fewer supers than there are those who, to some degree, are face blind. The spectrum allows for greater nuance in studying the degrees of face blindness, or indeed face recognition generally. Rather than something one either has or doesn't, one has a varying degree of ability. This is somewhat normalizing for those on the extreme ends, framing face recognition like any other ability or skill that has population variance, but in practical terms, it is the diagnosis that matters, not the fact that it lies on a spectrum.[19] Naming matters. It calls associations into being.[20] It creates categories of affiliation and identity.[21] It creates a concrete index that grounds a set of behaviors in an explanatory (and communicable) frame. Naming can be reassuring. Both ends of the spectrum have struggled to identify their face recognition abilities as being extreme, and, indeed, diagnosable.

The diagnostic history of super recognition teaches us a great deal about how we identify and name conditions, and what we consider problems or abilities. While the diagnostic tools for face blindness and super recognition were quite similar in the first instance, it took sixty-two years and four individuals to self-identify after the clinical classification of prosopagnosia for the latter to be identified following and absolutely as a result of the naming of the former. Even though both ends of the spectrum were hard for people to note in themselves, the one that caused significant social challenges was studied significantly earlier, as were the challenges themselves. The recognition of faces mobilizes a whole set of relationships around identity, reputational capital, and the very nature of being human. Living in a sea of blank faces may have its advantages; it certainly has much to teach us about not judging others by how they look. And living in a world of infinitely variable faces can teach the same thing: when every feature is fundamentally unique and unforgettable, they all become equally meaningful . . . or meaningless.

The world of the super recognizer is a world of exquisite and sometimes overwhelming detail, in which strangers glimpsed for a moment are established in memory equally as much as someone encountered every day. Prosopagnosiacs live in a world of virtually indistinguishable, or perhaps blank, faces, in some cases unable to differentiate by sight their mother or lover or best friend or even themselves from a stranger. Oliver Sacks recounted that "I've sometimes had the experience of apologizing to someone, and realizing it's a mirror."[22] On both sides there is a lot of social deception, but with a huge difference: super recognizers learn to pretend they don't recognize people, and prosopagnosiacs try to pretend they *do* recognize people.

But the face blind can't always fake it, unlike the supers. It's a lot easier for the supers to get away with pretending not to remember fleeting encounters than for the face blind to succeed at pretending to recognize their nearest and dearest. For most, it's hard to be face blind; it's hard to be in public with face blindness, and it's even hard to be in private with others with face blindness. Supers also live in a sea of visual information that is detached from intimacy or emotion. For them, like the face blind, face recognition alone is a meaningless indicator of prior relationships and connections. Super recognizer and West Midlands police officer Andrew Pope has said that it is "very hard to switch off when I'm off duty. I'm constantly looking at faces when out with family and friends."[23] While this may be an occupational hazard generally, it's certainly more acute when the chances of recognizing someone in the crowd are so much higher. As we'll see, the supers tend to think the cognitive load is worth it; it may be, and also they have no other way of being in the world. Face-blind people like Dori Frame, whom we met in chapter 2 and who talked about the "sea of strangers," may seek out crowds in order to avoid the pressure of recognition; for the supers, that is simply never possible. There are no visual strangers, just faces that immediately become indelible in their brains. This is, for the supers, the normal way of being in the world, just as never recognizing anyone is standard for the face blind.

Seeing across a Spectrum

A few key super recognizers are featured across many of the news articles and television shows. Their perspective is important, but so is that of people who have not yet been identified or diagnosed. To understand how *they* describe and articulate their experiences of super recognition before and while it is identified, I turn to a Reddit thread in which many people discover that they may indeed be supers; this thread provides valuable insight into what issues they consider as they process this information. Their discussion of their past experiences and interests going forward are incredibly similar, underscoring themes of social awkwardness and monetizability as they figure out that their creepy knack has been studied and named.

Jennifer Jarett, one of the original four super recognizers, echoed the lack of need for deep connections in her super ability, saying that "I never forget a face. I don't need an emotional attachment to someone to recognize them."[24] London police super recognizer Andrew Eyles expressed the same sentiment in different terms to the *Washington Post*, emphasizing that from a recognition perspective, loved ones and strangers register exactly the same: "I might only deal with (someone) once. But it would be like looking at a family member or friend."[25]

There are social and relational implications to these equivalences. We don't know what they are and they are not being examined, because the stakes seem to be quite low: supers are doing just fine. Current research is more interested in identifying the age at which super recognition is established; looking at eye-scanning patterns; and scanning the brain to see activity during recognition.[26] But supers are not simply the highest end of the face recognition spectrum, with just this one advantage. Rather, like the face blind, super recognizers have developed a whole host of accompanying or compensatory mechanisms to deal with their extraordinary ability. And they may well have deficits that have arisen from not having to have strong relationships with people in order to recognize them. Alongside the insights around labeling we can learn from examining the rhetoric around super recognition, there is much to be learned from studying how supers form relationships, how they consume media, and what their experience of recognition is like. That's not lab science, and it can't be explored through scanning the brain. But it has important implications for the role of the face in human communication and interaction. Studying together the way both extremes form emotional connections could play a significant role in unsettling the primacy of the face. It would also reframe spectrum models by considering challenges, abilities, and adaptations together to understand exactly what the face does *not* tell a super. There has been a great deal of exploration as to what the face means to a face-blind person, and the answer is very little. That is, in a way, also true of supers.

I'm IMDb: I Must Be a Super

Supers are super at something rather useful, in a rather specific context.[27] And the word *super* clinches it: that which is super must be useful.

It's not so easy to match a picture to a person. Prior to the establishment of the super squad, the United Kingdom's CCTV didn't result in many apprehensions: experienced criminals hid their faces, avoided cameras, or simply could not be identified due to the poor quality of the images. Face recognition software didn't work most of the time, and when it did produce a result, it was often subject to significant racial and gender bias, as we discuss in chapter 6. There are also logistical challenges to making identifications from images: there are a lot of people out there, and not all their details are on record with the police. And the images are not always clear, especially in the case of professionals who know how to obscure or turn their faces to avoid the cameras.[28] People in face-matching jobs—customs and immigration agents, police officers—perform around as well as the general population in matching faces to photos, which is to say, not very well. Numerous studies have found that error rates tend to be between 10 and 30 percent.[29] Face recognition software doesn't always do much better; it is prone to significant programmer bias, particularly around race. Face-matching systems do very well with mug shot images, which are taken under ideal conditions with full frontal shots and good lighting; other kinds of photos, which may have poor lighting, unusual expressions and body positions, and strange angles, had historically and until recently been much more difficult to match to a person or other photo.[30] Supers, some scholars theorize, tend to focus on one feature, and so are unaffected by these error factors: as we'll see, they used to do much better than face recognition software in correlating London's poor-quality CCTV images to both live people and mug shots.[31] Hey: it's a job.

But what's a non-British super to do?

Outside of London, there just isn't a huge market for super recognition beyond appearing on game shows and news spots, and occasionally being borrowed by various police forces.[32] It is a cool skill, and over 3 million people to date have taken the test to find out if they belong in the super category.[33] Many suspected that their abilities were outside the norm already, but absent an obvious application, there seemed no point or path to finding out if they were really (or clinically) unusual. The stories that supers tell about their recognition experiences are all eerily similar: they talk about the moments of recognition they eventually learn are found creepy and they eventually learn to suppress, the ability to extrapolate and age people over time (also, it turns out, experienced as weird), the really quite satisfying ever-ready answer to the question: where have I seen that actor before? To learn more about their stories, I turned to a Reddit AMA (ask me anything) thread with Dr. Ashok Jansari, a Goldsmiths, University of London, professor and super recognition researcher.[34] It is one of the only sites as of now where super recognizers have consolidated to discuss their abilities, and there is little to no communication between supers themselves; instead, they mostly ask questions of Jansari. The supers have not (yet) formed around an identity category; they don't need to for advocacy, treatment, or support. That may be changing slightly with attempts to monetize their abilities: supers are starting to find each other to help find jobs—for example, with Davis or Super Recognisers International, largely in security positions.

There was a great deal of similarity among the reactions of posters to Jansari's explanations: so-called symptoms of recognition came up repeatedly with people who thought they might be supers.[35] I was particularly interested in the experiences of those who did not yet have the label of super recognition alongside already identified supers as a way to understand if the label itself matters beyond an initial sense of relief and enlightenment. It turns out that no one quite knows what to do with the information or the ability, though it would certainly be useful for politicians, clergy, fundraisers, and basically anyone whose job depends on building relationships and making people feel special.

Nixpolosion asked, "What are some of the 'symptoms,' for lack of a better word, of being a 'Super Recognizer' that you've been able to ID?" He continued to list his own indicative experiences, noting that "Maybe Im reading too much into it but I may be/have [super recognition]. Ive been able to, for a long time, recognize people I havent seen in years . . . or recognize say, an actor who was a background/one off character in a film I saw a long time ago and be able to point them out in another film where they are also a backgrounded/side character." Jansari responded that these were in fact classic signs, saying, "You are 'full of symptoms'! Yes, those are basically the things that people tell us—that they recognise people that they only knew as kids, that an actor who has a bit part in the background of a scene is recognized, that they meet someone very very fleetingly and then remember them totally out of context a few years later." Anecdotes aren't enough though: while Jansari acknowledged that "you sound like a super :-)," the doctor made sure to ask, "Have you done my test yet?"[36] Jansari always needs more supers to study; other researchers also need supers to staff their growing security and face recognition businesses.

But it isn't clear, or wasn't to people with these abilities, just what the value of super recognition is. Reddit user FlandersFlannigan wrote about being "extremely gifted with recognizing faces" that "friends and family have even recognized this about me and they call it my useless superpower." Not only "because it [is] very much useless" but it is "even embarrassing sometimes." FlandersFlannigan explained that "I can't tell you how many times, where I'll talk to someone and be like 'Hey [name], we briefly met 10 years ago while you were out on a morning jog.'" Responses to this kind of interaction are as most of us non-supers would predict: "9/10 times I'm met with the *creeped out face*." Now that FlandersFlannigan learned that "this is a thing," which "I'd never thought," their next reaction after "This is so weird" was to try to figure out how, exactly, to make the superpower a little less useless by asking, "What do people with this ability typically do for work?"[37]

Super recognizer Moira Jones would also very much like to know the answer to that question, writing in a piece for *The Psychologist* that "if anyone reading this has any thoughts on how, or where I might apply this skill, please do let me know." She had only recently realized that there may be specific possibilities and potential because it is only "now [that] I know I have a very specific talent, I am eager to put it to good use."[38]

Reddit poster Afgangsta was also relieved to discover that their ability had a name, writing that "Ive always recognised people that I've met once in my life and then seen them again after, lets say 5–7 years later." But recognition is only one part of what others experienced as creepiness: "Not only do i remember their face i remember the conversation we had, if we had any." And, again, "99% of the time they are freaked out as to why i remember, so i tell them i have a good memory but they still look at me weird." To avoid freaking people out and being looked at weird, "now even if i run into someone i don't say that i remember you incase they think I'm a stalker or something." Here, Afgangsta reflects that people expect to have some kind of connection with those who recognize them; supers are outliers in not needing one. Like FlandersFlannigan, Afgangsta was relieved to have a category for what to them was natural but to others might seem stalkerish: "Its great to know that there is a name for this ability: Super-Human. Nice."[39] Indeed.

Nixplosion, FlandersFlannigan, and Afgangsta's descriptions of their abilities and the challenges of people's reactions to them echo across the thread. Many, like Afgangsta and others we've met, describe developing coping mechanisms to avoid making others uncomfortable, which is something they'd had to learn over time, not initially realizing that their recognition abilities were out of the norm. Redditor iamambience wrote that "Its nice to never forget peoples faces, but I learned early in life that 'regular' people find it weird that you can remember them if you only met them once before, and time has passed in between." Here, iamambience noted two key symptoms of super recognition: the recognition itself after only brief interaction, and the retention of that ability over time, which requires not just recalling the face in question, but aging it. Again, reactions were wary, as "They read too much into it, and think your last encounter meant more to you because you didn't forget them."[40] Again, face recognition is, for most, tied to emotion and connection; for the face blind and supers, it isn't.

iamambience referred explicitly to the ability to recall faces without any meaningful emotional connection. For most people, recognition is the result at least in part of some kind of memorable or unusual or powerful aspect to the encounter, particularly if it is singular and fleeting. For supers, this is simply not true: they do not need any kind of tie, deep or otherwise, for someone's face to be seared into their brain. That can, as iamambience pointed out, also be awkward. They recall a "funny story: Over the past 13 years I have on four occasions run into a stranger I once talked with on a bus. I freak him the fuck out every time."[41] Though by the fourth time, the stranger might remember iamambience because of being so freaked out, an example of the kind of strong emotion that supers don't need to remember and then recognize faces but often elicit in others.

For Redditor eunonymouse, there were professional disadvantages to the ability. They "worked in retail and would frighten customers by remembering them from months ago." As a way to cope, "I now usually pretend not to remember people, makes life easier." Lamzn6 agreed, saying that "I'm a super recognizer and I have the same experience. I have to pretend like I don't recognize people." PmknSharkLatte chimed in that "This happens to me as well!" It's uncomfortable all around, and "I feel bad though because I find that most of them don't remember me," so "it might be better to just pretend that I don't recall meeting them."[42]

The hard parts of being a super almost never come up on TV.

That's not an accident.

It's Totally Worth It (But Is It?): The Rhetorical Framing of Superness

Part of the positive perception of super recognition has to do with the use of supers in crime detection. It's a great narrative: super skill lets super sleuths track down bad guys, triumphing where the technology fails. It's also not false: when chief of Scotland Yard Mick Neville distributed CCTV pictures to his force back in 2007 (a full two years before super recognition was clinically identified and tested), some of his officers immediately made more than fifteen identifications from the grainy pictures.[43] Super recognizer officers identified 197 people from CCTV following the 2011 London riots, many of whom were only partially visible due to face masks, poor photo quality, and partial or profile pictures.[44] The strength of super recognizer identifications as evidence is so efficacious that with the evidence unearthed after the identifications are made, three out of four cases reach prosecution, as opposed to the one in five rate for traditional apprehensions.[45] The seemingly unsolvable case of the murder of 14-year-old Alice Gross in 2014 had a breakthrough when a police super recognizer was able to correlate a person visually captured biking on a tow path with someone caught on camera in totally different clothing buying alcohol at an off-license, a store that sells alcohol to be consumed elsewhere (as opposed to, say, a pub).[46] The super squad also played a key role in the recent identification of the Sergei and Yulia Skripal poisonings in Salisbury in March 2018.[47] While the bulk of super recognizer identifications happen in London, other international cities are now paying attention and even borrowing these unique officers. Following the train station assaults of women in Cologne in 2011, German police worked with the supers from London to identify the assailant and the victims from a large number of suspects visible in the video footage.[48] Supersecurity is going global: researcher Josh Davis is currently working with a number of international police forces and criminal justice agencies to set up super recognizer units in various contexts. As we learned in chapter 3, he has created new diagnostic tests to account for different racial configurations across the world.

As a way to catch criminals, or, more specifically, as a way to suture the shortcomings of CCTV, the super recognizer unit (formally established in 2015) works.[49] There are a lot of caveats, and it is important to emphasize the difference between matching video to photos and databases and memory within controlled conditions as opposed to doing so on the fly. Super recognizer–aided identifications are not themselves admissible in court, but they do provide additional opportunities and options in any given case. As we have seen, several researchers are actively exploring the differences between super recognition in controlled conditions and outside the lab, and the results, they emphasize, should deeply temper our expectations and perhaps even the role of supers within criminal justice.[50] Most super squad members work on identification with, but not exclusively with, CCTV. As the face recognition technology improves in accuracy and reach (for better and for worse), it will always need people to make sure it is doing its job right. Even if the face recognition technology improves, as long as there are cameras, there will be jobs for some of these supers.

And if Gary Collins, a super with London's police force profiled by the *New York Times* in 2015 "almost was punched" now and again because "I think sometimes I stare a bit too long, but I can't help it," it's probably a decent trade-off for being "off the charts," so extraordinary at recognizing faces that he excels at his job beyond all expectation. And if "he deliberately lives outside London to avoid running into wanted faces from his beat," or has to "cut short an outing to the mall with his sons when he recognized a gaggle of gang members while buying sneakers," it's worth it for what he calls "the gift." Like other supers, it was hard for Collins to name his ability. As he said, "I always recognized people, but as kid you don't know . . . you just think everybody is like you."[51]

But of course, almost nobody is like the supers, as far as we know. The clinical understanding of super recognition is still very limited, and often changing. In Jansari's words, "The work on super-recognition is EXTREMELY new in research terms—the first paper was published in 2009 which in research terms is barely an eyeblink!"[52] There is much to be learned, from the technical and neurological questions around visual processing and the fusiform area of the brain, to the psychosocial and political implications of these abilities to cross-racial and cross-gender visual recognition.[53] There is *also* much to be learned from how supers form relationships, and how they understand faces encountered both interpersonally and through media.

Super recognition is of course a useful skill, even with the awkward (but not debilitating) social dynamics that supers have to learn to navigate. There are practical implications in terms of where supers live, what they do professionally, and how they navigate public space. There may be other sorts of challenges as well. So too may super recognizers have some arenas they never had to or may even not be able to develop. In the Reddit thread with Jansari, bavarian_creme wondered if indeed such deficits exist, recognizing the simplistic narrative around super recognition and probing further. They stated, "Of course super-recognition sounds like a super power, but I imagine it's not that simple." bavarian_creme got more specific, asking Jansari if "you learned something about the negative effects super-recognisers could experience due to their way of perceiving other people?" In particular, "of psychological or social nature, that maybe even affects their day-to-day lives in some way?"[54]

The short answer Jansari gave was essentially "no." The longer version was more like "not yet." The very long version was that it was a good question, in that "in some areas of research where people have an 'ability' that others don't have, we look to see if this comes at the 'cost' of something else." Jansari explained that for people with synesthesia, the "enhanced skills in SOME areas" may have led to them being correspondingly "weaker in others." But "since super-recognition is at a very early stage of research, it is difficult to comment whether there are any weaknesses associated with the ability at the moment."[55]

We like our superheroes to be invincible, with just a tragic flaw or two to make their stories all the more impressive.[56] Just as we like our disabled heroes to be supercrips, and all others to be villains.[57] But there is a loss here in this binary: both supers and face blind have unique ways of understanding faces and identities. Let's examine these ways together.

I Just Play One on TV

Supers have drawn a lot of media attention, including a series of television spots from news magazines like *60 Minutes* to the Fox game show *Superhuman*, among others.[58] These media outlets are highly diverse in nature, and yet they almost all followed a remarkably similar format in telling the story of the supers.

The model is designed to maximize the impact of this extraordinary skill, even resorting to trickery and manipulation. More baldly, as Jennifer Jarett explained in chapter 3, "it's faked."[59] The goal of media attention is to maximize the superhuman skill even at the expense of veracity. These shows all ask supers to identify people they have interacted with briefly, or even not at all, only passing them in the street or catching them out of the corner of their eyes. The supers are told at the outset that they are meant to find a face in the crowd, either on the streets or in a studio. They are given just a few moments to look at the faces or wander around the spaces, and are then asked to locate specific people. But there is always, always a trick: either the targets add a hat or change their clothes, or perhaps audience members change their seats while new people are added to the group as others are removed.[60] The trickery raises the stakes from showcasing a merely impressive skill to something truly superhuman. Something totally worthwhile and something totally positive. Something that should be applied in daily life as soon as possible, ideally in service of making the world a better and safer place. That's what superheroes do. To underscore the larger picture, commentators and hosts always talk about there being few or no places for people to hide because *the supers never forget a face*. The targets are civilians, but the message is clear: criminals, beware. Surveillance isn't just from cameras.[61]

Everyone is always floored. The hosts, the audience, the targets: without fail, they all express deep admiration and shock that the super can do precisely what she was supposed to do. The shock is also part of the structure, to emphasize for those watching that this isn't a trick that the show itself is part of, and that the skills are genuine. And genuinely impressive, if even jaded television personalities can be affected so much. The follow-up is quite vague, however. In absence of a clear career path or application for super skills (outside of London or with the security placement firm Super Recognisers International), these television shows, much like the supers themselves, don't know quite what the next steps should be. There isn't an obvious professional or practical application, so the shows present parlor tricks with an underlying narrative around superheroes, leaving it to the audience to connect the dots.

Little or no attention is paid to any potential personal challenges the supers have as a result of their abilities beyond briefly mentioning the moments of social awkwardness caused by recognizing people they don't know.

This small humanizing detail serves to acknowledge that there are social implications to the ability that may not be ideal but are still entirely manageable and ultimately not really harmful. Unlike for people with face blindness, the social awkwardness that supers encounter is quirky and even at times astonishing in its own right, and in any event, it can be quickly overcome. A little social awkwardness is nothing a superhero can't deal with on her own. The powers of her mind overcome the challenges presented by the demands of the body that cannot be eliminated. Indeed, embodiment and its ramifications remain a key component of being a superhero, no matter how cerebral or digitized or mechanized her skills.[62]

Some super recognizers are starting to find each other and share their stories, like in the above Redddit thread. But not many: unlike face-blind people, they don't need support and community in the same way around daily coping mechanisms, strategies for watching television and movies, and the basic sense that they are not alone.[63] As a diagnostic category or community, they also haven't been around as long, and, partly because of that fact, they don't have as much research and material to share. There's a growing amount of both scientific and humanistic literature on prosopagnosia, but the field of super recognition is much newer.[64] A number of labs continue publishing their findings, but they include no humanistic analyses of the politics, experience, and social consequences of super recognition. Much as we can learn from the face blind the importance of paying attention to identification cues beyond the facial, so too can we learn from the supers other implications of vexing the face recognition–relationship index.

We know, or at least we are starting to know, that super recognition can be detected even in teens, which tells us that at least some facial recognition cognition develops earlier than previously thought.[65] What we've also known for a long time is that most people recognize others based on emotional connection, personal interaction, and repeated encounter.[66] That sort of recognition is not true on either end of the spectrum, though; face-blind people recognize almost no one, regardless of prior relationships, and supers recognize everyone. We know a lot about the challenges and costs for people with prosopagnosia.[67] We know almost none of the challenges and costs for the supers. Super recognition is a superpower, and we aren't quite as concerned about the challenges of superpowers if we can use them to save the world.

Conclusion: Supercreep?

Supers present the opportunity for an easily narrativized story of people doing something better than machines (face recognition software can't even come close, nor can biometric kits) through a really cool new application of a previously under-recognized and still under-capitalized set of abilities. It's not because supers are more intuitive than machines; it's that they are looking in a fundamentally different way, focusing on specific features and identifying them across platforms. At the moment, face recognition software can't do that. And if super recognizers haven't quite managed to apply these skills outside of CCTV crime detection and private security, they soon will. In the meantime, we are still finding out who the supers are. So are they; thanks to easily accessed internet tests combined with the media coverage, more and more supers are being identified, studied, and (in London) put to work every day. The ones who aren't being put to work (that is, most of them) are honing their abilities in daily life and on TV, preparing to save the world, or at least identifying extras on television. (Some kinds of social awkwardness are more useful and potentially monetizable than others.)

Like prosopagnosiacs Oliver Sacks, Jo Livingtone, and Ben Dubrovsky, whom we met in chapter 2, it's a process for many supers to figure out that they can put a name to their condition—a newly recognized diagnosis with even newer online diagnostic tools. It's hard to reflect that face recognition isn't just a matter of having a good or bad memory, that other people have a magical ability or are just a bit bad at something seemingly basic.[68] And at both ends of the spectrum, there are social consequences: much like those with face blindness, super recognizers tend to have socially awkward encounters resulting from the way they recognize others. As we've seen, that's in fact one of the so-called symptoms of these conditions. Super Sallie noted that "I often have to lie that I've never met or seen people before" because "it freaks them out" and she doesn't want to "sound like a stalker." Amanda too said that she has "given up on" telling people where she recognized them from as she "got too many weird looks and I started to feel like I was creepy sometimes."[69] But the social and personal stakes are different on both ends: not recognizing loved ones is bad. Recognizing everyone, even those with whom there is no personal or temporal connection, while creepy, is good. While supers all report awkward encounters with people whom they recognize after only brief meetings, with whom they have no emotional connection but whom they recall as though they were dear friends, they also learn pretty quickly to stop mentioning it or to pretend they don't recognize others. They learn to strategically deny their superpowers. Again, that's not that hard, and even kind of charming as a set of social challenges. The ability to recognize every face ever seen is good. It's amazing. It's super.

That's how we talk about superpowers. And this is one: it's right there in the name. Even if, as super Andrew Pope said, "it felt a bit strange, to be honest, as the word 'super' is quite overwhelming," it must be positive.[70] And it must be of use for saving the world in some way, because that's what superheroes do. That's what the rhetoric around super recognizers is designed to emphasize: that catching criminals on CCTV will save the world, even if not all the supers are doing it yet. Can super recognition save CCTV? As we'll see, face recognition technology is rapidly improving, though despite both positive and dystopian hype, it will not ever operate entirely autonomously. Computers do not think. They are not intelligent. They will not be able to destroy us or save themselves. But they can and do track us—effectively in some cases, less so in others. In chapters 5 and 6 we will look at the long history of surveillance, computing, and face recognition. Just as CCTV did not invent face recognition, it also did not invent surveillance. We've been tracking people for the purpose of sorting who is in and who is out for quite some time. We're getting more granular in our methods, but I would not say that means we are getting better. Even as the technology improves, perhaps that means we are getting worse.

5

Face Surveillance at the Border

Checkpoint Charlie

Face surveillance and its faults did not begin with face recognition software—not even close. It's embedded in much broader systems of racial capitalism and oppression, medical taxonomy, human genealogies, the sorting and prioritizing of certain kinds of people at borders and those who are allowed to cross them, and even portrait painting and visual culture. It's a computing system but also a biometric one, inseparable from the history of colonialism, racial violence, and the hierarchy of human types. The narrative of security carries with it always the question of whom we are patrolling against; on a global scale, we can see caste cameras in India, facial surveillance of the Uyghurs in China, and indeed attendance surveillance in US schools with low-income and majority-Black populations.[1] The story of face surveillance lies not just in the history of computing but in the history of medicine, race, psychology and neuroscience, and in the health humanities and politics.

Face recognition is a local story about relationships between individual people, but it is also a global one. We have explored how super recognition has been leveraged for policing and surveillance in the twenty-first century, but there is a longer history of deciding based on faces who is in and who is out. Here I discuss analogue attempts to make the people in charge of gatekeeping and borders better at face recognition and therefore better at surveillance. The face has long been used for identification not just for personal relationships but for administration, national identification, border patrols, and policing writ large. In all these uses, there have been gaps, shortcomings, biases, mistakes, and indeed failures. As the story of face blindness shows, failure is an integral part of making recognition itself visible. There is a mean-

ingful difference between individual face recognition and the systemic attempt to track faces. Impersonal systems built on face recognition are
designed to keep track of and differentiate between kinds of people. Face
recognition software is only the most recent manifestation of global systems
of tracking and sorting. Here, we excavate a missing link in the long history
of systems of face recognition that emphasize what we already knew: it's always been about surveillance. And in this case, it's about the most surveilled border of its time, Berlin's Checkpoint Charlie, and the head border
guard, Peter Bochmann, who invented face recognition testing just as much
as the prosopagnosia researchers we have introduced. And he invented it for
specific reasons: to limit passage between East and West Germany; to lower
the number of errors and mistakes on the part of the border guards who too
frequently allowed people to cross with fraudulent or stolen passports; and
to improve the practice of matching people to pictures, all in order to increase
the number of people caught.

Bochmann was very good at his job. Some of the techniques he developed
look a lot like the ones that face recognition researchers use today. As I excavate this piece of history, I reluctantly elevate the work of the head border
guard of a deeply repressive secret police. There are stakes to highlighting
how people did bad better. Many historians have begun to reflect on the ethics of telling the secret stories of others; while these are important issues to
discuss, if we do not tell the secret histories, we continue only with the official ones, written by those with power and as they wished the narrative
to proceed. My goal in telling Bochmann's story is both to highlight the
broader implications of the long history of biometric technology and to
emphasize the role that surveillance has always played in tracking people
by their faces and bodies. In so doing, we see the deep links between the Cold
War context and the development of face recognition technology. While the
mechanisms may have gotten more sophisticated, the goals have, in many
ways, remained the same.

It's the Same Story the Crow Told Me: The Greatest Hits of Biometric Surveillance

The Wende Museum in Culver City, California, preserves Cold War
artifacts and culture and has done an extraordinary exhibit on Checkpoint
Charlie border inspections.[2] The museum has also told the face recognition
and biometrics story through a rapid narrative linking figures like Lavater,

Lombroso, Galton, and Bertillon, and moving quickly up to Checkpoint Charlie, Ekman, and modern face recognition technology.[3] All of these (mostly male) figures deserve their own section and indeed books, and many authors have offered precisely that attention.[4] They are the figures most often cited in the literature about face blindness, creating a standard historical narrative that has become a kind of orthodoxy.

Swiss pastor Johann Caspar Lavater (1741–1801) created a particular system of the age-old practice of physiognomy (the relationship between physical features and character) that had long seen use in the arts, divination, theology, philosophy, and, for a short time in the late eighteenth and nineteenth centuries, systemic scientific study. Lavater offered one approach to judging people based on their features; Franz Joseph Gall (1758–1828) and Johann Spurzheim (1176–1832) drew from Lavater to establish his system of phrenology, which was another approach to linking the physical body to character and action. Gall and Spurzheim's model was deeply intertwined with the rising middle class and the self-improvement ethos of the mid-Victorian era. He proposed a series of organs of behavior and character that were expressed as head bumps; the larger the bump, the more prominent the behavior. Those whose organ of, say, charity, was too small could exercise and grow it. Unlike physiognomy, which had a fixity inherent in its approach, phrenology was more fluid in its indexicality to character. Both were forms of somatic recognition, one more visual and one more tactile. They served as early technologies of bodily identification, offering systems that claimed to know who people were based on how they looked or how their bodies were expressed.

Italian psychologist Cesare Lombroso (1835–1909) built upon the somatic fixity of physiognomy and combined it with elements of criminology to argue for a physical marker of criminality. More specifically, Lombroso's method found criminal behavior to be inherited and marked upon the body by visible atavistic traits that clearly identified degenerates to the trained eye. He offered detailed measurements and research to support his biological determinism, elements of which lingered in the criminal insanity plea and the establishment of institutions for the criminally insane. French police officer Alphonse Bertillon (1853–1914) worked explicitly with bodily measurements to provide a system of identification and record keeping that could track criminals following their release. Bertillon's system was cumbersome and generated an impractical amount of data, but the principles were refined with the later incorporation of fingerprinting technology appropriated from Sir

William Herschel's (1738–1822) application of the practice in India to the United Kingdom by Francis Galton (1822–1911) in 1892. Galton, perhaps best known for developing the principles of eugenics, also offered a model of composite photographs. He combined images of many different portrait subjects through repeated limited exposure to produce an ideal of a given type. Galton's goal was to produce a definitive visual representation of, say, "the criminal type" or "the sick type" for future use and identification purposes.[5] While his technique failed to prove common and recognizable features across any single type and instead demonstrated that a group of portraits would tend toward an average, the notion of somatic links between face and character continues to persist, sometimes in highly creative and interesting ways.[6] Contemporary artist and coder Jason Salavon, for example, has created a series of images that average frequently photographed moments in order to expose common or even universal features, conventions, and practices. He has argued that his work excavates meaning from the unexamined features of shared humanity, though he and others working in this medium acknowledge the potential surveillance and criminal justice applications to the process.[7]

The "great men of somatic identification" narrative starting with Lavater usually peters out around Galton and his eugenic practices, possibly drawing some links between these modes of surveillance and contemporary examples. This story is a carefully plotted and tightly chronologized series of developments focused on Western male scientists with deep investments in physically identifying deviants while cultivating an equally easily representable elite (white) class. Such approaches, some have argued, found their logical culmination in the eugenic and genocidal practices of the Nazi regime, with other versions persisting globally.[8] These practices include the legally supported reproductive eugenic sterilizations enshrined in the United States in the *Buck v. Bell* case of 1927, which allowed involuntary sterilization of the mentally ill or physically disabled, and which was upheld through the end of the 1970s.[9] More recently, the California prison system authorized forced sterilization of around 150 inmates between 2006 and 2010; other states have similar records, as does the Immigration and Customs Enforcement (ICE) agency.[10]

While eugenics is one clear trajectory or application of these behavioral somatization attempts, another narrative follows the story to face recognition software and modern surveillance techniques, particularly regarding the use of AI for diagnoses of disability.[11] (And, of course, these two endpoints—eugenics and surveillance—are themselves intertwined in many ways; there are those who are granted social privilege and basic rights, including, at the extremes, reproduction and freedom, and those who are not.) Surveillance is a way to detect the deviants according to whatever rules might be in place, be they: the laws of a democratic society; the genocidal frameworks of dictators; or, as is most often the case, a tiered set of rules that include the enforcement of systems of structural repression for some and access to rights and privileges for others.

This particular narrative of biometric identification, which we see almost everywhere face reading is featured, entails a leap of anywhere between forty and seventy years. After the intellectual parade of white men whom history records as being the innovators of research on faces, there is a shift in focus to building on this work to chart the steady development of eugenics and face recognition techniques respectively. These are key historical moments, part of a steady and nuanced progression that is rooted in broader cultural, social, and geopolitical developments. But even in these more finely tuned studies, some clustering of research occurs: Lavater to Galton, then eugenics and racism in practice, then face recognition software. (Woodrow Bledsoe, credited with founding computer face recognition, may or may not get a mention; he's been rediscovered in recent years.)[12] There are of course many other relevant moments in that forty-to-seventy-year gap between the Nazis and Clearview AI, or between the Nazis and eigenfaces (one of the major innovations in face recognition technology), or even between the Nazis and Identi-Kit technology.[13] Long before people were trying to teach machines to recognize faces (and correct their mistakes), people were teaching people how to recognize faces. While, as we've seen, most people do okay at recognizing faces, some people do better than others. But the people whose job it is to recognize faces, and in particular to recognize faces in one context and compare them to faces captured in another, don't actually perform much differently than anyone else. Specifically, border guards, the bulk of whose job is to compare the people in front of them to their passport photos to determine if they do indeed match, do not necessarily possess particularly strong abilities in doing so. They are, for the most part, not super recognizers, though a few may happen to be. Aside from some small tweaks on the margins, they, like everyone else, can't really improve, at least as far as we know to date.[14]

But that doesn't mean that they don't try. And it doesn't mean that people haven't been trying to improve face matching in this specific arena for a long time. Face recognition software and AI is one such attempt that we'll explore in chapter 6; here we look at an analogue effort at improving face recognition that is often overlooked. Programs designed to draw on the skill of super recognizers to train border guards proliferate to this day.[15] Results are modest at best. Stakes are high. But this sort of training isn't new: border guards have been tested on these skills for a long time, especially at one of the most historically heavily surveilled borders in the world: Checkpoint Charlie.

At Checkpoint Charlie: The ABCs of Face Fraud Detection Training

The Berlin Wall was built in 1961, creating a physical and tangible separation between East and West Germany in particular, and the East and West in the Global North more broadly. The wall made passage between the two halves of Germany more challenging, but like all borders, it remained porous. The Grenzübergangsstelle Friedrichstraße-Zimmerstraße, known in English as Checkpoint Charlie (figure 5.1), was the only designated and allowable border crossing point for both members of the allied forces and non-German citizens.[16] The checkpoint was highly visible and gained considerable cultural traction in film, television, and fictional narratives.[17] In its way, Checkpoint Charlie has remained quite famous, now serving as a major tourist destination with many of its artifacts displayed in museums worldwide. Checkpoint Charlie Mauer Museum was founded shortly after the construction of the wall and has been operating since October 19, 1962. The checkpoint was busy, both with legitimate crossings and those seeking to flee to the west or to access loved ones (and cheaper shopping) in the east. As Marieke Drost has chronicled in "Nose Matching at Checkpoint Charlie," the first three years of the wall saw a huge exodus of East Germans using borrowed passports from West German citizens who resembled them enough to pass the possibly substandard scrutiny of the border guards. So the border guards got tougher, or at least their training did: in 1964, guards staffing the border between East and West Germany fell under the administration of the notorious Stasi, who developed extensive instruction guides and testing mechanisms to improve the detection of identity and passport fraud.[18]

The Wende Museum in Culver City has an extensive collection of training manuals as part of its Peter Bochmann collection. The manuals include

Figure 5.1. Checkpoint Charlie, April 8, 2013. Courtesy of Jaipreet Virdi

detailed images of faces and side-by-side comparisons of features designed to improve verification techniques and limit the number of people getting through with stolen or borrowed identity documents. Broken down by feature types for eyes, lips, nose, and head shape, profile, nature of facial hair, and the triangle of nose, lips, and chin, these manuals guided guards through specific types of comparisons that highlighted how to categorize features and see if they correspond between person and picture. In Marieke Drost's evocative framing, this collection of "biometric technology *avant la lettre*," took the underlying theories of greatest hits guys like Lombroso and Bertillon and applied it directly to border patrol, producing a small industry of not just material but testing mechanisms, with the ability to perform well on matching tasks now part of the requirements for the job.[19] If nothing else, given the challenges in improving matching skills, the presence of the Stasi ensured a more rigorous checking process that would itself eliminate any errors born of lack of close scrutiny. It seemed to work: by 1975, almost no one with a fraudulent passport slipped through the checkpoint.[20]

Almost no one. But as with any barrier, people will find a way through. In addition to the underground networks that had people traveling literally underground or through unauthorized breaks in the wall, a scant few still managed to borrow or steal identity documents and pass directly through the checkpoints. It was only when the theft of these documents was reported (after the fact) that the guards discovered they had erred, and by then the fraudster had long since crossed and, practically speaking, disappeared. Which meant, to the newly appointed head of the East German passport division (known only by his first name, Peter, in his initial interview with Drost in order to protect his identity from the contemporary social repercussions of having participated in the Stasi), people were not doing their jobs well enough. Note that "Peter" is GDR border guard Major Peter Bochmann, who gave his holdings to form the Bochmann collection at the Wende Museum. Bochmann remains protective of his identity in German-language publications, but is willing to be on record in English-language contexts, and he has allowed a video interview to be posted on the Wende Museum website as part of their "Facing the Wall" exhibit.[21] In the video, Bochmann alludes to how porous the border could be, particularly for someone in his position, noting that "I always had the opportunity to leave the GDR illegally by stepping 1 metre further." "But," he continued, "I never even considered that" because "this is my homeland" and "I loved my country."[22]

Bochmann was stationed at Checkpoint Charlie for long stretches of a career that started in 1964, and he served there almost exclusively from 1975 until the Wall fell in 1989. As he rose through the ranks, he continued to observe people illegally passing from East to West. If some were getting through, there was room for improvement in the training of border guards and their skills at matching and detection of identity fraud. This, in Bochmann's view, was a solvable problem. A specialist in facial recognition, Bochmann had focused on the topic in his Stasi college in Potsdam, and he brought his skills to improving the outcomes at the border, especially his home base of Checkpoint Charlie. As Bochmann recounted to Marieke Drost, he was ideologically committed to the GDR, but that wasn't what motivated him to begin his research into how face matching worked at the East German border. He hated when jobs were done poorly, particularly when the application of science and data could improve outcomes, though the challenging work conditions may also have contributed to error. In his video interview, Bochmann described working at Checkpoint Charlie "around the clock from the early shift to the late shift," for "9–10 hours on the short shifts and 13 hours on weekends or holidays." They were granted "4 days off" only after "14 days," but "often had to come in then as well because of a shortage of people."[23] It was, suffice to say, a demanding job at an intensely busy border.

After observing several cases of successful fraudulent crossings in his first two years in his job, in 1977 Bochmann began an extensive research project designed to improve outcomes at the border. Rather than focusing on improving working conditions to reduce error, he spent time analyzing how experienced guards examined faces and what features they would focus on as they made their comparisons. Bochmann determined which of these methods seemed most successful and used this approach to develop a training module for new recruits built upon the work of their more experienced colleagues. He developed flash cards with two photos on each; new recruits had to successfully determine whether they matched as part of Bochmann's exacting training program to improve how well guards matched people to passport photos. These flashcards bear a striking resemblance to tests currently used to diagnose super recognition and to more broadly determine face recognition accuracy; as we'll see, Bochmann's research anticipated the direction that later neuroscience would take with respect to face recognition, which at that time had still only identified a scant few prosopagnosiacs and no super recognizers at all.

The relationship between face recognition and border patrol is an intimate one that was galvanized by the introduction of photos into passports in some countries in 1914.[24] Much of the early face recognition software research was funded by the CIA for the purposes of border surveillance.[25] The CIA built on this research to develop a standardized framework for face segmentation that emerged from police evidencing and spycraft practices around surveillance and patrolling.

Seeing Like a Recognizer

Bochmann's archive is rich in images and descriptions. He put his guards through a rigorous program of facial recognition, drilling them in different categories of lips, eyes, head shape, crown shape, and profile. Figure 5.2 shows pages from a Stasi training manual (*Merkmale des Äußeren von Personen* [External distinguishing features of persons]), to train border guards in specific feature types, attempting to improve recognition and attention to differences. The manual included multiple facial categories, including the position of the eyeballs, the mouth, hairline, hairstyle, facial hair, and lips. The distinctions were quite granular for facial hair, differentiating between a moustache that was "thin, thick, toothbrush," or "upturned, straight whiskers, hanging." Similarly detailed categories appeared for beards, with a caveat that "sideburns are to be noted." The manual included examples of eyebrows that were "busy" or "sparse," "shaped by shaving," and "penciled in." The "position of the eyeballs" urged guards to note if they were "deep," "protruding," or "crossed" in various ways including "inward on one side," "outward on one side," or "inward on both sides." The mouth and lips also included detailed notes, ranging from size, width, and how and where the lips were "jutting."

Figure 5.3, titled "Exercise 6," shows the application of the techniques presented in figure 5.2, asking trainees if each of the two depicted image pairs represents the same person, respectively. As part of their response, trainees were asked to "determine the identity/non-identity of the persons portrayed in the image pairs 1–5," and "provide comparisons of 4 distinguishing characteristics in support of your assessment." Trainees were then directed to "please turn page" to continue practicing.

Figure 5.4 is a series of flash cards that Bochmann developed to test matching skills, while figure 5.5 depicts a recent super recognizer test (the Glasgow Face Matching Test); note the similarity to Bochmann's exercises, though the modern test does not ask for proof of the claim.

In addition to his taxonomy and its grounded application, Bochmann developed a broader theory of successful facial recognition based on breaking down the face into its component parts and focusing on specific features rather than the whole when matching to photos. Bochmann's taxonomy had analogues in later literature, which to date is deeply engaged in understanding holistic rather than featural processing in both the general population and those at the extremes of face recognition.[26] While the data remain inconclusive as to whether super recognizers tend more to featural processing, many suspect that those who excel at face matching in particular tend to employ this technique, likely instinctively.[27]

Featural processing cannot, to the best of our knowledge, be taught.

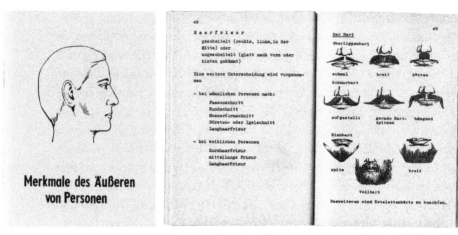

Figure 5.2. *Above and opposite.* Cover and excerpts from *Merkmale des Äußeren von Personen* (External distinguishing features of persons), a manual for training Stasi border guards in face recognition. From the Peter Bochmann Collection, courtesy of the Wende Museum

Nevertheless, Bochmann and others in the Stasi tried. And perhaps they improved outcomes at the margins, increasing care and attention and highlighting specific techniques to determine if a set of features matched between photo and person despite different hairstyles, facial hair, posture, or makeup. Figure 5.6 shows more granular notes for Stasi guard facial recognition training, drawing attention to particular facial characteristics, profile types, and face shape.

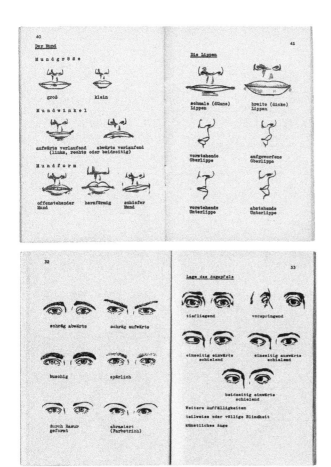

It is powerful to note how Bochmann's work resonates with current facial recognition technology, not only in how the data were compared (because in this case people's features were considered data: that is, they were used to turn identification into a problem to be solved) but how they were collected. As Bochmann admitted, he began secretly taking photos of travelers and their passports. It was tough to capture both, and only about one in twenty came out clearly enough to be usable. But that was enough for Bochmann to develop another module to train guards in the matching of faces to pictures based on a sample of those actually seeking to cross the border. As they discovered, even among the limited sample of one usable photo of every

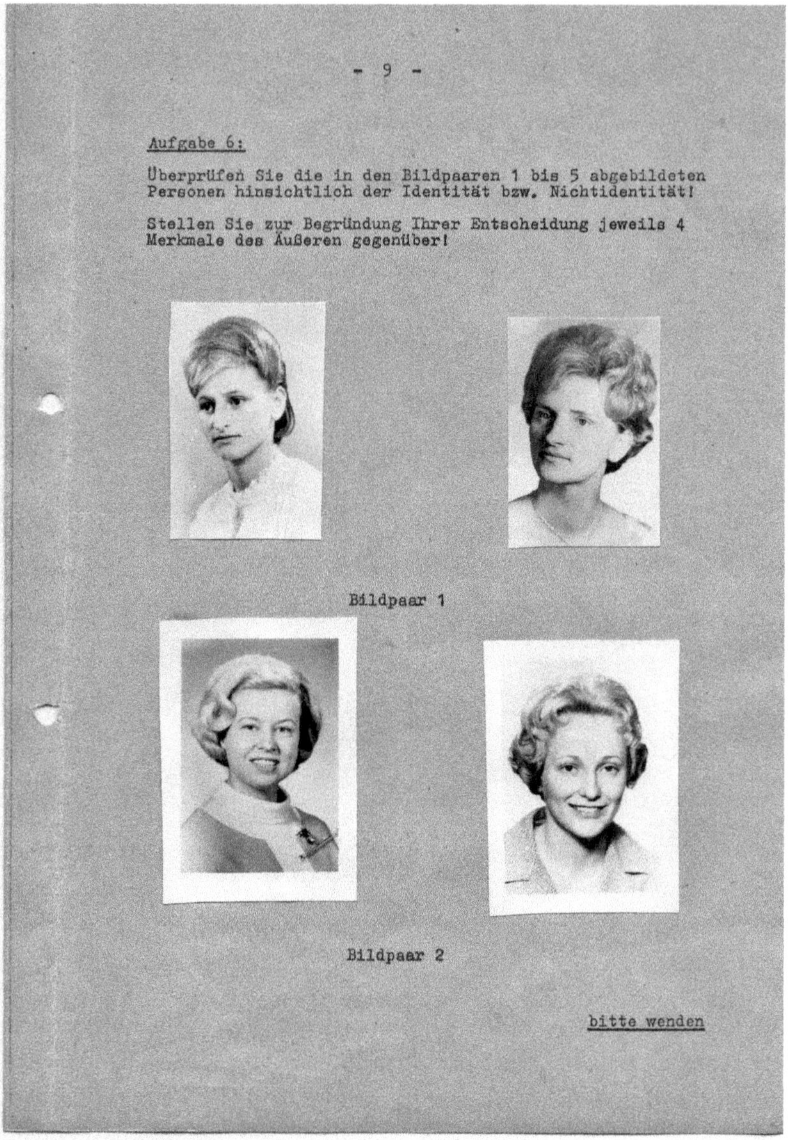

- 9 -

Aufgabe 6:

Überprüfen Sie die in den Bildpaaren 1 bis 5 abgebildeten Personen hinsichtlich der Identität bzw. Nichtidentität!

Stellen Sie zur Begründung Ihrer Entscheidung jeweils 4 Merkmale des Äußeren gegenüber!

Bildpaar 1

Bildpaar 2

bitte wenden

Figure 5.3. A face-matching test that Peter Bochmman gave to trainees. From the Peter Bochmann Collection, courtesy of the Wende Museum

Figure 5.4. A set of flashcards that Bochmann developed to test face-matching skills. From the Peter Bochmann Collection, courtesy of the Wende Museum

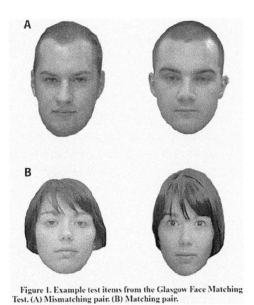

Figure 1. Example test items from the Glasgow Face Matching Test. (A) Mismatching pair. (B) Matching pair.

Figure 5.5. The Glasgow Face Matching Test. Pair A: mismatching. Pair B: matching. Courtesy of A. Mike Burton

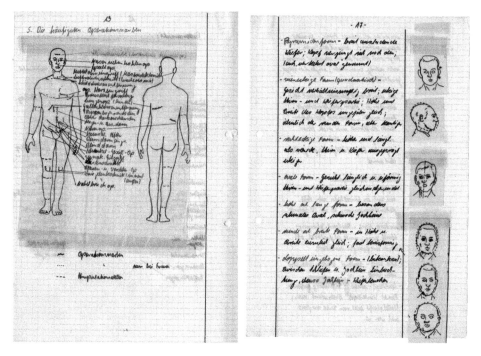

Figure 5.6. *Above and opposite.* Notes from *Merkmale des Äußeren von Personen* with specific attention paid to face shape and profiles. From the Peter Bochmann Collection, courtesy of the Wende Museum

twenty, there were examples of fraud. Discovered too late in those instances, but perhaps preventing fraud in the next.[28]

Bochmann recognized that there were multiple ways that a passport could be fraudulent; in addition to stolen or borrowed documents, travelers could prepare unofficial or forged papers. He compiled an extensive set of descriptions of the features of various national passports from his secret photos, providing his border guards with key characteristics of each country's documents in hopes of spotting faked papers. Figure 5.7 provides an example of one of Bochmann's descriptions, outlining the minutae of a Finnish passport. The notations discussed the cover type, material, color, type of imprint, and color of imprint. The descriptions went into detail about the binding, the text colors, the printing process, the watermark, and the UV light, even noting that the color of the embedded security fibers, visible only under fluorescent light, is "none."

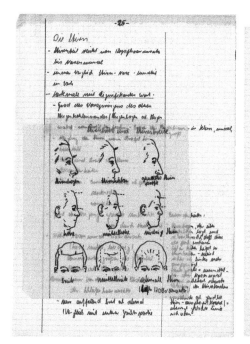

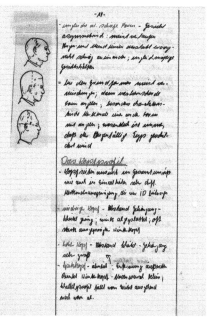

Peter Bochmann and those like him who were quietly toiling away in border surveillance and contributing to the long tradition of bodily taxonomy and facial recognition are not part of the greatest hits narrative. Bochmann didn't publish his work, and only very recently did he discuss his findings and his system. His reticence to speak was practical: he did not want to be known for contributing to a deeply oppressive regime, even though he largely supported its work and aims. (He wished he could travel, he said, but the trade-offs were worth it.)[29] My choice of Bochmann as a link in the chain is not accidental. It is not simply a fortuitous archival find thanks to the generosity of historian Paul Lerner, who alerted me to the Wende Museum and introduced me to curator Joes Segal and writer Marieke Drost; rather, it is a deliberate choice to bring the Cold War into the conversation. This historical moment provided the conditions of possibility not just for one of many high-stakes border crossings, but the weaponization of data in the ongoing global theater. For Bochmann and others at Checkpoint Charlie, data was a way to more effectively keep people in. For those hoping to develop better mechanisms for machine translation, it was, as we'll see in chapter 6, a way to listen in.[30]

Paßbeschreibung

Land: **Finnland**   Paßart: **Reisepaß**

Paß Nr.: **0103943**

Reg. Nr.: _____

ausgestellt am: **2.12.87**   in: **Helsinki**

Einband

Art: Decke, <u>Umschlag</u>

Material: Plast, Kunstleder, Leder, Leinen, Pappe, Karton

Farbe des Einbandmaterials: **dunkelblau**

Art des Aufdruckes: Pressung, Reliefprägung, Druckverf.: _____

Farbe des Aufdruckes: **goldfarben**

Broschurblock

Bindungsart: F-genäht, F-geheftet, geklammert Plast verschweißt, <u>Rückstich</u>, seitlich

Farbe des Textaufdruckes: **braun**   , ~~schwarz~~ XXXX

Druckverfahren: Offset-, <u>Buch</u>-, Tiefdruck _____

Wasserzeichen: hell, dunkel, <u>Relief</u>, ohne

Farbe der Melierfasern: _____ , <u>ohne</u>

UV-Licht

Fluoreszierender Untergrundaufdruck: ja, <u>nein</u>.

Fluoreszierender Textaufdruck: <u>ja</u>, nein

Latenter Untergrundaufdruck:
(nur im UV-Licht sichtbar)   ja, nein

Farbe der fluoreszierenden Fasern: **hellblau**   , ohne

Fluoreszierender Heftfaden: <u>ja</u>, nein

Besonderheiten

Foliierung auf Seite: **Spiegel**

Güst **Fri-/Zi-Str.**

Datum **24.12.87**

Mitarbeiter **Fw.Polster**

Zutreffendes unterstreichen bzw. eintragen.
Für notwendige Ergänzungen Rückseite benutzen.

4 - 11 - 2

Research That Helps to Do Bad Better?

Should Bochmann and the Stasi border-guard training find their way into the historical record around face recognition research? Given the impressive similarity between Bochmann's work and the Glasgow Face Matching Test developed some forty years later, it is tempting to wonder what direction Bochmann's work might have taken were it more mainstream. It is equally tempting to sideline this work, given that it is inextricable from the uses to which it was put. The Stasi was one of the most repressive and effective secret police forces of the modern era. It is not difficult to make a case against the Stasi, but that is not the focus of this project or my engagement with Bochmann.

It is easy to condemn the Stasi surveillance project.[31] I introduce Bochmann and his work both to offer more continuity in the development of face recognition technology which is, in its way, an effort to make people into supers, and to highlight the gaps in any teleological or "greatest hits" approach to historical narrative. Bochmann is one of many people who continued to refine face recognition technology. Had he not given his archive in pieces over time, which itself was the result of Bochmann's simply taking all the documents home when Checkpoint Charlie closed in 1990, we might never have learned of his work.[32]

Bochmann's relationship with the Wende Museum developed due to hard work, happenstance, and luck, slowly and over several years. As with so many archival donations, the curators had to gain Bochmann's trust and develop a relationship with him; given the sensitive nature of these historical events and his concern for his own safety and reputation, the process was delicate. Wende librarian John Ahouse conducted the original interview that appears on the Wende website; over time, Bochmann came to feel comfortable with the intentions of the museum and gave his collection to them in stages. Bochmann was particularly concerned to ensure that his material would not end up in a German institution that would only emphasize totalitarianism rather than a more nuanced historical perspective around the history of border control in the GDR.[33] But it is worth noting that Bochmann's documents proudly bear his name ("Peter Bochmann Border Guard Collection") and that his own video testimony includes no apology for his research or activities.[34]

It's the historian's job to excavate not just the information and outcomes but the broader process and cultural and social narratives in which that information was produced precisely in order to frame it in its fraught context, tricky as that might be. This is the work: to unsettle the narratives as we know them, partially because they are what we know. I seek not to use Peter Bochmann's research to improve surveillance, create more supers, or help us do bad better, but rather to situate a clear historical instance of the dangers of this kind of project. While border patrols continue globally in both democratic and non-democratic states alike, and granting the possible dangers to passport fraud, the Checkpoint Charlie case offers one of many grounded examples of the insidious applications of surveillance. In addition to the algorithmic bias inherent in current face recognition protocols, a broader question exists regarding the role of the state and indeed the role of capitalist infrastructures in developing ever better ways to know who, what, and where we are at any given moment—and selling that improved information, or trading it, or acting upon it to change our behavior or punish it.[35]

Given the ongoing nature of these questions in light of the fragility of many global democracies, their urgency will only grow. Writing about those who have done research is not the same thing as using or endorsing that research; I am not interested in evaluating the accuracy of Bochmann's technologies or applying them; rather, I seek to know how they may be consonant with other work in this arena, including current approaches that are also inextricable from the surveillance they enable.[36]

Conclusion: Somatic Surveillance to Serve the State

Somatic surveillance cuts multiple ways: originally designed as a tool of identification and negotiation, it was quickly applied to the criminal justice realm for purposes of detection and incarceration. The standard narrative of face and body recognition has often cut a clear swath through some key historical figures; in this chapter I've discussed a gap in the record that is likely one of many examples of innovations in facial recognition designed to hone the craft and improve results. Designed, indeed, to make people better at face recognition. Designed to improve analogue systems of surveillance to better control the movement of people across borders. What emerges from including such examples is that facial recognition—whether done by people or, as we'll see, computer programs—is imperfect. There's a limit to how good any one individual can get at face recognition and how much improvement can be achieved: most of us will never be super recognizers, and most of us won't get better than we already are. But as Bochmann's detailed interventions and surprisingly extant archive show, people continue to try. In Bochmann's case, people illegally left (and entered) East Germany under a totalitarian regime; the Cold War context is central to galvanization of surveillance research.

As we'll see in chapter 6, both early machine translation efforts and face recognition technology emerged in the Cold War as ways to look and listen in. Some current practices in face recognition have deep resonance with Bochmann's work, though the mechanization of detection is a point of departure. The role of surveillance, and the sorting of people into those who can move freely across space, remains constant and connects current face recognition software directly to Bochmann's work in limiting passage across Checkpoint Charlie. The stakes remain high: recent applications of facial recognition include not only border patrolling but false positives leading to perpetuation of racist tropes, false arrests, identity error, and the upending of lives. There is much to be concerned about with respect to face recognition technology, be it human or automated. Mistakes are made. But even as the algorithms improve, these practices remain a fundamental incursion upon individual liberty and privacy that is only increasing as the practices improve and the technology becomes more ubiquitous. People's level of discomfort varies on that front, but it is clear that face recognition practices are at odds with privacy by design, and the practices, from early biometric surveillance through Cold War innovations and to today, have ever been thus.

Face Recognition Software and Machine Translation

Why Computers Aren't People

Machine learning is a misnomer. Machines don't learn; they compare. And through comparison, they refine. The term "artificial intelligence" is likewise an artifact of the early days of computing, when machines were designed to mimic the approaches of humans. The name stuck, but while machines are artificial, they are not, in the way this complicated and historically contingent term is often understood, intelligent.[1] Training and learning are not the same thing. Intelligence and comparison are not the same thing. But this language, which elides the ways that computers are designed to get the idealized output by the best mechanism necessary, which is often decidedly not the human one, has persisted. Maybe because we trust learning and intelligence, the implication of a teleological process is a comforting one.

Is face recognition technology the new and better super recognition system? Where Bochmann failed in his analogue project, has the digital version succeeded? Here, we look at the intertwined Cold War histories of face recognition technology and machine translation as attempts to mimic the work of the human brain in order to do a better job of surveillance. The history of the technology is fundamental to understanding its goals and development; these are built into both the process and the product. Super recognition systems were and are specifically designed to track people—across borders, in commerce, and in law enforcement. And these once sluggish systems are good and, increasingly, better. The error rates go down as the speed goes up.

Not so with people: Jennifer Jarett has rejected all applications of her super skill beyond small talk, partly because of the limitations of what she can do but largely because when it comes to surveillance, she simply does not want to do it. Surveillance is, for supers, a choice. Declining rates of false positives don't mean that AI is rendering supers obsolete, nor do they mean that the problems with the technology are disappearing. We'll explore here *how* machines are different from, and utterly reliant on, people (and their biases). The things that people and computers do might seem to look the same from the outside, but the internal (invisible) processes are very different. That means that you can't learn about how people think by the way computers calculate; when a computer recognizes faces, what it is really doing is comparing a set of data taken from one problem—a particular face—and comparing it to great numbers, maybe even millions, of other collections of data, which is to say other faces. It then tries to eliminate as many errors as possible in determining if these two pieces of data are the same. Sometimes it successfully eliminates enough error to get it right. Sometimes it does not. It initially needs a person to help it know if it has done so in order to know which errors to eliminate in the future; with enough starting examples, it makes better comparisons that refine (or compound) over time. The word "errors" here does a lot of work: there are mistakes in the comparisons, but many of them emerge from the starting biases of the data sets and those who input them. If, for example, there is only one Black man in the initial set of faces for comparison, the machines will incorrectly assert a match with multiple faces that share only the most basic features, in this case Blackness and (presumed) maleness. While this "algorithmic bias" has been the subject of numerous scholarly works and even films of late, it continues to permeate technological infrastructures and it continues, for those not directly affected in the first instance, to be invisible.[2] But even if these biases are reduced to the point of insignificance, the entire infrastructure of these bodily surveillance technologies, as Os Keyes has argued, remains part of a broader set of oppressive and racist systems.[3]

We turn now to the overlapping histories of machine translation (MT) and face recognition technology (FRT) as Cold War infrastructures from the 1960s through the 1990s to understand the roots of face surveillance and the distinction between what our human brains do and what machines do when recognizing faces. People on all points of the face recognition spectrum remain integral to any face recognition machine infrastructure. Peter Bochmann's system was one kind of Cold War surveillance technology; machine translation and face recognition software emerged from the context in which Bochmann was working. I tell the parts of the story that relate specifically to the history of the face recognition spectrum and how we have learned how people recognize faces. And how we have not, which is equally relevant when it comes to the misleadingly named machine "learning" at play in current face recognition technology.

We will also consider contemporary implications: face recognition software is part of how super recognition is currently being monetized and applied, and together these elements are working to create a new surveillance infrastructure. These systems are embedded in the history of machine learning and human-computer interaction, and, as we'll see, face recognition technology is always going to be intimately connected to and reliant on human face perception, because that's how deep machine learning works. Increasing accuracy and the growing scale of these systems came to obscure the human role even as these developments lay bare the biases. Improvements should not be credited to the machines actually learning, at least not in the way that learning works in people. Again: machines do not think like people. Machines are not people; people are not machines. They both do some things well, but they are not the same things. To put a finer point on it, we cannot learn from machines how people think, or translate languages, or recognize faces. A great deal of AI funding from the 1970s was predicated on precisely the attempt to replicate neural networks for the purposes of learning how people work, or, more specifically, to eliminate hypothesis and determine how their brains did *not* work. Reproducing what humans do—without agency or choice—was the means rather than the end.[4] It was only when FRT researchers like MIT's Alex "Sandy" Pentland shifted away from this model toward programming recognition for its own sake that it began to work.[5] Even though a machine can recognize a face and a person can recognize a face and it looks like they are doing the same thing because the outcome—a recognized face—is the same, the process is not. Because machines are not people and people are not machines.

Neither computers nor people could ever work at enough scale, speed, and accuracy to surveil a city; together, they complement one another's strengths. The story of face recognition software is not the story of successfully replacing humans with machines. It was also never the story of learning about humans from machines. Instead, much like the history of machine translation, the story of face recognition technology is the story of how humans become part of the machines. In the process, we confuse the function for what is inside the (black) box. Face recognition technology compares large data sets. Human face recognition recognizes faces. The latter trains the former to make fewer errors. But machines cannot teach us how people recognize faces, nor can they do it better. They can just do it more.

Mechanism is useful in this case more for what it can't teach us—how human face recognition works—than for what it can, which is how machines compare data sets. FRT started to work at scale when programmers shifted their focus from trying to mimic human cognition to trying to maximize output. For face recognition software, this most fundamental tenet of how people make sense of one another, build relationships over time, create bonds between others and express their own identities on and through their bodies, has become a question of calculation. For most people, recognizing faces is a whole other thing connected not just to neurological function but to relationships, history, and experiences. For the machines, recognizing faces is math.

People and computers can look at two images of faces and decide if they are the same. But people recognize faces and computers compare data. Which means that we can't learn how people recognize faces by learning how computers do it, because recognizing faces isn't what computers do. To understand where computers go wrong, we need to truly appreciate what it is they are doing, and why it is they are improving. The short answer is that they have bigger data sets and better memory. The long answer is that people are helping them get better at eliminating errors. The even longer answer that we'll explore here lies in the story of machine translation, which is not about faces at all. But it is about how computers learn, and the necessary role that people play in helping them. In both machine translation and face recognition software, there is an early history of trying to get machines to mimic the cognition of people, which may well have shed some insight into how the human brain works. Except that it didn't work at all, and instead, both processes developed as (different) methods to compare large data sets. And they got better at doing the thing that people could already do. They were just doing it in an entirely different way. They did, and still do, depend entirely on the human element to get better and maybe even get it right. I'll use computer language learning as a model for why face recognition technology will only ever be as good (or bad) as its programmers, and what that means for face recognition more broadly. We'll see how face recognition technology originally tried to mimic what people did. That didn't work. So researchers pivoted and tried to solve the problem a different way. Let's pause and reflect on this framing, which was widespread in the literature: rather than acting as an ethical conundrum or a social mediator or a surveillance tool, face recognition technology was a problem to be solved.

The Lives of Others

Face recognition surveillance and machine translation intersect in the field of state control generally, and the Cold War in particular. As Michael Gordin and others have outlined, the history of machine translation, and, specifically, the history of trying to make machine translation viable and effective, can be situated in rising hostilities between the Soviet Union and the United States beginning in the mid-1950s.[6] Gordin's fascinating account of the history of Soviet machine translation is a historiographic intervention that considered, among other things, the unit of the algorithm itself, the notion of algorithmic reasoning, and their role in the construction of stories of dominance, progress, and technologic prowess. While Gordin has kept a firm eye on technical details, his focus is simultaneously detail-oriented and big-picture, showing that what has been excluded in the historical story of MT is just as, or more, illuminating than what historians have tended to emphasize.[7] For our purposes, Gordin has already done the work of linking global geopolitics with the history of computer-based state hostilities; his focus was on US-Soviet relations, but the implications for the East German border patrol then, and face-based surveillance now, remain deeply resonant. Also, Gordin's insistence on the algorithm, and its reliance on people to make it work, was precisely the point in the eventual success of machine translation in the 1960s, following a period of fevered research that Gordin has called the "gold rush of 1954–66," and others have located in the launch of Sputnik in 1957.[8] And the algorithm was also precisely the point in the eventual success of face recognition software in the 1990s.

There are a number of unsurprising common threads across the origins of machine translation and face recognition technology that are actually not that different after all; aside from their origins and applications in governmental surveillance, we see early attempts to use machine processes to mimic and thus learn about how people think. We also see these attempts being part of the problem rather than the solution for the question of how to get computers to translate language and recognize faces. It is only when these questions were reimagined as computational rather than cognitive that programmers could get the machines to do the work. At first, programmers tried to encode grammar into the machines to have them read things the way people read things. But it's hard to map out grammatical structures of language as a heuristic; people can do it quicker and cheaper than machines. Instead, it made sense to take a large already-translated corpus and let the machines compare and go from there. Memory got much cheaper over time, making this method so efficient and inexpensive that it almost felt like thinking. But it wasn't. It did, however, work. Just in a different way than people do.

Success required recognizing that the goal was the output, not the process. And this could only happen, as Gordin outlined, when the computer programmers, technologists, and mathematicians put people squarely into the process. In a way, that's obvious: machine learning requires people to identify errors in the comparison of data sets: without people, the errors only compound. Consider machine translation: given that language evolves over time, a translation program needs people to update it. Without people in the process, language translation will ultimately remain static or even descend into gibberish for long periods of time, regardless of how robust the starting data set it. (And the Canadian in me must note that the French-English translations of Canadian parliamentary procedure in *Hansard's* have played and continue to play a key role in machine translation, being among the most comprehensive starting data set for comparison in those languages.[9]) But the human role, as Gordin wrote, is "often missing in that classical understanding of algorithm." As the history of both MT and FRT teach us, "all algorithms require embedding in a person before they can be realized in a machine." This is clearer than ever in today's technological context for both language and faces; in both cases, "deep learning has brought us back to his [Russian Machine Translation innovator who lived from 1894 to 1950 Petr Smirno-Trianskii's] insight, for a human determines that the neural net is making matches correctly." But, as Gordin insisted, despite the historical record, "this seemingly discarded technology reminds us of a vision of algorithms, and of language, that have always been with us."[10]

Stephanie Dick's account of the 1956 Summer Research Project on Artificial Intelligence at Dartmouth College is another discussion of the misnaming of AI. Dick historicized the ways that human cognition is fundamentally different from computer cognition; while the outputs may be broadly similar, they are doing very different things to get there.[11] She recounted the original approach of those at the workshop, which was designed to "identify the formal processes that constituted intelligent human behavior in medical diagnosis, chess, mathematics, language processing, and so on, in hopes of reproducing that behavior by automated means." But it did not stay that way, in large part because it simply didn't work. Instead, "artificial intelligence today resembles this symbolic approach in name only." The original goals shifted from trying to learn about these processes at work in human intelligence in order to reproduce and automate it in machines.[12]

Contemporary efforts focus almost exclusively on designing systems that work regardless of whether they do so in human ways.[13] Many of today's complex problem-solving systems exclude human behavior entirely; game-playing programs, for example, hone their strategies by competing against themselves instead of against people or through heuristics based on human play, noting successful strategies and developing often unexpected ways of winning.[14] Dick here referred to the system's methods of comparing data sets of winning moves, an approach that is entirely automated and bypasses human strategies. Even in these cases, however, people design the systems. The algorithm did not build itself, but the mechanism progressed in entirely non-human ways, leading, as Dick has described in particular with the game Go, to novel strategies and surprising results. Both Gordin and Dick have discussed when and how this transition from focusing on process (mimicking people) to product (maximizing output) occurred.[15]

The programmers took a people process and made it into a math process. Although it is a different process it often gets mischaracterized as the same one. Data anthropologist Ali Alkhatib recently discussed the persistence of this error. He introduced the notion of a "problem space arbitrage" in which programmers attempt to solve human problems through technical solutions. They acknowledge the trade-off but imagine the payoff of working in the technical format to make it worth it in terms of progress. Alkhatib's work is attentive to both the new problems that this penalty introduces, and to the ways that the programmers become the arbiters and experts entrusted with solving human challenges through machine workings. Not only does this shift the locus of authority around who tries to solve these problems; it determines who gets to define them.[16]

The early literature on face recognition software development, when for the most part it didn't work, describes a similar trajectory. Initial approaches from the 1960s through the1990s tried to make the machines recognize faces the way people did, or rather the way researchers at that time thought people did. Programmer David Silver, who in 2017 helped create the grandmaster-level Go-playing program that defeated champions, noted in a 2020 interview in *Wired* that "we haven't built something which is like a human brain that can adapt to all these things." Rather, "the real world is massively complex"; the latest version of the program, MuZero, "really is discovering for itself how to build a model," which it is learning "just from first principles." Silver knows that the term "first principles" is doing a lot of work here; even in this largely independent system "we tell MuZero the legal moves in each situation" from which it engages in reinforcement learning. Interviewer Will Knight pushes the point, asking "Some observers point out that MuZero, AlphaGo, and AlphaZero don't really start from scratch. They use algorithms crafted by clever humans to learn how to perform a particular task. Does this miss the point?" This question in some ways gets to the heart of what an algorithm really does, and what sense we make of those who craft it. Silver responds emphatically that yes, people are missing the point, because while "you never truly have a blank slate," and "there's even a theorem in machine learning—the no-free-lunch theorem—that says you have to start with something or you don't get anywhere" this case is different. Indeed, he insists, "the slate is as blank as it gets." Except, of course, "we're providing it with a neural network, and the neural network has to figure out for itself, just from the feedback of the wins and losses in games or the score, how to understand the world."[17]

The distinction Silver made here is between systems that need a human interlocutor to approve or exclude matches between data sets, as in FRT or MT, and those—like MuZero—that determine from the outcome (winning or losing) what the best moves are. Ultimately, these kinds of systems have different goals, so I agree with Silver that the question about algorithms misses the point, but not because the algorithm is somehow more independent in MuZero than in the other cases, but rather because the system itself is closed; whether a "move" or a comparison between two data points offers the correct outcome is determined by the game itself. Of course, we still need a starting algorithm, but also the game is comparing data points between all the moves played in the past. The outcomes have already been decided, or, in the terms of the other systems, whether a comparison is correct is the equivalent of the outcome of the move. It still matters how it turns out, but the mechanism that determines the outcome and its goals is different. And the difference, in a way, is what also made MT and FRT finally work: seeing these systems as independent structures that work very differently from the human brain. The machines got much better, but not because they were getting more human in their processes. Because they never were human. Once the problem is framed as one of comparing volumes of data derived from, say, human features, rather than mimicking human reasoning, it could be solved. It was only a matter of time.

I mean that in two ways: it took time to figure out how to solve these problems, but also, it took the ability to compress time and data to make these things work at a scale that people never could. Or, indeed, that people alone never could. Super recognizer researcher Josh Davis, whom we met in chapter 4, succinctly outlined the inextricable relationship between people and machines, noting that "humans see faces in a human way, computers see faces in a computer way." These are distinct processes; "they aren't working the same way. There are occasions when humans make mistakes on faces that no computer would, and occasions when computers make mistakes on faces that no human would." Things work best when you have both approaches in tandem, such that "if a computer makes a decision and a super recognizer looks, chances are they will make the same decision. If you combine the two, you get the highest levels of accuracy."[18] Not only do computers need people, Davis argued, but people, even super recognizers, need computers—if the goal is to surveil the populace as accurately as possible.

There are several points of departure from the original plan to hack the workings of the brain by imitating it to focusing only on outcome instead of process. And of course, these points are themselves just links in a long chain that are utterly dependent on what came before and entirely necessary for what followed. For FRT, we can point to the 1990s and the use of eigenfaces, essentially a mathematical formula that produced standardized facial components that got more specific with more data points, that offered an algorithm to find basic characteristics that were common among some faces and not among others.[19] These templates can then compare the set of eigenface data points across two images to see if they match. Using principal component analysis, we can think of this holistic approach as reducing faces to data points and simplifying them to see what might match.[20] This was a powerful technique that achieved significant results, but it still relied on analyzing the face as a whole, mimicking the way that most people make sense of faces and identify them. Following the holistic approach of the 1990s, programmers moved to handcrafted local descriptors in the 2000s, introducing local-feature learning by the end of the decade.[21]

While I want to be careful not to get too technical in ways that are unhelpful or not germane to this book, it is a key claim of this chapter that we should unblack the FRT box and look inside so we don't misinterpret what it can (and cannot) do, and indeed what it is (and is not) doing. While these technologies are distinct from the human brain, they are fundamental to how we understand face recognition generally, and at its most accurate extreme.[22] Face recognition technology works by comparing captured images (photos or videos) with those in a database; it is programmed to determine if there is a match using a highly refined mathematical set of data comparison criteria. This programming is based on initial examples determined by people who verify when matches occur to give the machines a starting framework. The machine uses these examples to refine the basis of comparison and thus "learns" to improve and refine results over time. Originally formulated by computer scientist Woodrow Bledsoe in 1959, the technology has improved significantly over time, with a resulting increase in accuracy.[23] Bledsoe's original model started with a mechanism that recognized alphanumeric characters; the model was then extended to photographs of faces.[24] Japanese researcher Takeo Kanade built on Bledsoe's machine with a computer program that automated the entire recognition process by extracting features from a photograph without additional human intervention.[25] Bledsoe's starting database was four hundred white men.[26] Subsequent databases were not significantly more diverse.

Error rates, including false positives, have decreased dramatically over time through the various iterations of the technology, from software to hardware.[27] Early versions worked by taking a static image of the face and measuring the distance between features according to the starting algorithm. The results of this map would be compared to similar results in the database, which connected faces to names and identifying information. Later modifications tweaked the algorithm through component analysis, producing a more granular set of data points; current practices draw on the affordances of video capture, which improves upon pixelated static images by creating composite images and 3D models of the face and head from video to reduce the challenges of poor lighting and changing angles. The underlying principles of data mapping remain the same, but the capture is far more sophisticated. Pioneered on a mass scale at the 2001 Super Bowl to identify people in the crowd, including nineteen individuals alleged to have outstanding warrants, FRT is now basically ubiquitous in commercial and national contexts, from unlocking your phone to crossing borders to tracking your movements in the streets and at front doors.[28] These tweaks have considerably improved outcomes, bolstering claims of gradual improvement and increasing success.

A lot depends on the points of comparison. The more faces in the database the better for purposes of accuracy, though perhaps the worse for purposes of privacy, as the many lawsuits and regulatory debates about data scraping and use of facial images have highlighted.[29] That's a significant source of error: if the database has very few Black men (as indeed it did), there will be a much higher incidence of false positives in which the system will say that a captured image is a match for one of the very few Black men in the database given the lack of data points for comparison. The image in the database may be, say, a wanted felon, and any Black man captured of approximately similar body type and age range will then be a match, arrested, and subjected to the treatment meant for this other person, a wanted felon. It's happened. It happened to Amara Majeed in 2019, when she was falsely accused by Sri Lankan police of a terrorist bombing.[30] It happened to Robert Williams, who was arrested in Detroit in 2020 due to a mistaken match.[31] It happened when Amazon Rekognition's face recognition tool decided that twenty-eight Black members of Congress matched mug shots on file, given how few Black people were in its data set.[32] Face recognition technology carries serious consequences, and even if the starting data set improves, other biases will persist.

"Face Recognition: A Literature Survey" reviewed the state of the field in 2003, highlighting the importance of the shift from process to product, and underscoring the myopic perspective of some in the field with regard to ethics and application.[33] The essay considered FRT a cool and interesting math problem rather than a grounded technology with real-world political stakes. That attitude alone was a shift; the authors were thinking less in terms of mimicking human behavior and more in terms of surpassing it. It's possible they did not realize the potential applications; it's more likely they didn't care, focusing on the question rather than the implications of the answers. They emphasized the shortcomings of the technology in comparison to human capabilities but framed these gaps as a challenge to be met and exceeded once the affordances of the machines to work at scale were properly mobilized. The review was clear about the lacunae, noting that "recognition of face images acquired in an outdoor environment with changes in illumination and/or pose remains a largely unsolved problem. In other words," they clarified, "current systems are still far away from the capability of the human perception system."[34]

The authors did note the difference between working with people and working with machines, particularly on issues of consent and willingness to participate in data collection. FRT, they argued, was superior to other biometric approaches because "although very reliable methods of biometric personal identification exist, for example, fingerprint analysis and retinal or iris scans, these methods rely on the cooperation of the participants." FRT, however, can be used in stealth; the authors stress the advantages of "a personal identification system based on analysis of frontal or profile images of the face" as it "is often effective without the participant's cooperation or knowledge."[35] (It seems clear from this statement that the authors "didn't care" about the dangerous ethical stakes, not that they "didn't realize.") They doubled down on this framing later in the paper, arguing that "due to its user-friendly nature, face recognition will remain a powerful tool in spite of the existence of very reliable methods of biometric personal identification such as fingerprint analysis and iris scans."[36] Here we see not just a startling reflection on the perspective of these authors, but also a clear acknowledgment that further success in this field will best happen when the machines deviate from human processes. They made the importance of this shift explicit by stating that "it is futile to even attempt to develop a system using existing technology, which will mimic the remarkable face recognition ability of humans." Rather, the focus should be on building on the relative strengths of the machines, as "a key advantage of a computer system is its capacity to handle large numbers of face images," unlike "the human brain" which "has its limitations in the total number of persons that it can accurately 'remember.'"[37]

Toward the end of the essay, the authors repeated that "the impressive face recognition capability of the human perception system has one limitation: the number and types of faces that can be easily distinguished." People are good at this, but they can only do so much. "Machines, on the other hand, can store and potentially recognize as many people as necessary." They concluded with a rhetorical question, asking if it is "really possible that a machine can be built that mimics the human perceptual system without its limitations on number and types?"[38] The answer, of course, is no, but it *is*, they implied, possible to build a machine that surpasses the human perceptual system by doing it a different way.

The 2003 essay is dated from a research perspective, of course; it noted for example that "psychophysicists and neuroscientists have been concerned with issues such as whether face perception is a dedicated process (this issue is still being debated in the psychology community [Biederman and Kalocsai 1998; Ellis 1986; Gauthier et al. 1999; Gauthier and Logothetis 2000]) and whether it is done holistically or by local feature analysis."[39] While the question of holistic processing is still being debated, researchers today know a great deal more about the processes of face perception and face recognition.[40] This essay is valuable as a snapshot into not just what researchers were thinking about face recognition technology, but *how* they were thinking about it: in this case, as a problem to be solved, not necessarily by imitating the way that people do it.

This point was again reinforced as recently as 2019 in a post by scholar Jason Brownlee, who trains developers in machine learning systems and code. He noted that "face recognition . . . is a task that is trivially performed by humans, even under varying light and when faces are changed by age or obstructed with accessories and facial hair." While "trivially performed" might be an overstatement, particularly around changes in age, Brownlee's point is that recognizing faces is something most people just do, and something that most computers just can't; he underscored that "it . . . remained a challenging computer vision problem for decades until recently." However, once researchers moved away from a cognition model and toward a comparative one, the problem was mostly solved. Brownlee, whose career is based on deep learning methods, argued that "deep learning methods are able to leverage very large datasets of faces and learn rich and compact representations of faces," thereby allowing "modern models to first perform as well and later to outperform the face recognition capabilities of humans."[41]

The performance only exceeded that of humans by leveraging speed and the storage of massive amounts of information. This approach solved the algorithmic problem and rendered human faces as data points. Even as face recognition technology exceeded some humans, it did it in a different way. Which means that even if the people need computers, so too will the computers need people. But that makes the role of the supers less clear.

The technology has not eliminated errors, but it has come much closer to doing so, for better and for worse from a practical, ethical, legal, and social perspective. When the heart of a technology is people itself (is it ever not?), the ethical questions may work differently than those we tend to apply to machines. As the math became clearer from the 1990s through today, the ethics became a lot muddier. And ethics, as bioethicist Arthur Caplan has insisted, is method. Faulty ethics means faulty method means faulty data.[42] But these two problems are not distinct. In fact, as the work of the algorithmic-bias researchers has shown, they are in many ways the same. And that matters for the work of face recognition, and it also matters for how we conceive of the relationship between humans and machines, and humans as the redemptive technology for the shortcomings of machines. Which means that the biases of the humans get reproduced and amplified in the technologies of the machines. And it has always ever been thus.

That doesn't necessarily mean we should fear them.

Shiny Happy Machines

Science Studies scholar Lee Vinsel built on key insights from sociologist Judy Wajcman to warn cogently against what he calls "crit-hype"; don't, he cautioned, imagine that big tech (or AI, or Big Data, or whatever you want to call it) can, in and of itself, destroy the world.[43] It may, one day, but the dystopian capabilities assigned to technology far outstrip its current affordances. That doesn't mean that the capitalistic technology infrastructure (and the academic industries they generate) isn't hugely problematic—it is, of course—but, Vinsel emphasized, there are so many other pressing problems whose current trajectories will destroy us first. We should be thinking deeply about them instead of worrying about the latest shiny thing.[44]

Face recognition is, in its way, shiny, and its dangers have already been well realized and carefully documented. Various cities have banned its use, and three of the big tech giants have, after much pressure and well-publicized mistakes, declared moratoria on their own research and sales. Microsoft, IBM, and Google have all halted their face recognition programs, though of course Clearview AI has not. And it is capitalizing on that space. Australia and Canada have both engaged in lawsuits against Clearview for violating privacy laws that govern the use of facial images and their conversion to biometric data. Sweden has also investigated the unlawful use of Clearview tools by police forces.[45] These countries have more restrictive data privacy laws than the United States, with the exception of Illinois's Biometric Information Privacy Act; Clearview continues to fight the lawsuits brought against them in Illinois.[46] In addition to criminal justice applications, there are a host of attempts to mobilize FRT for identity verification; the IRS's use of the ID.me platform has so far proven to be a colossal failure that has prevented people from accessing their unemployment benefits and ability to make payments.[47] The company has attributed these failings to human error.[48] Insurance companies and medical start-ups are experimenting with FRT as a diagnostic tool, with significant concerns that these algorithms will be used to deny people coverage.[49]

The excitement around FRT's capabilities is perhaps matched only by the critiques: many people think FRT is going to revolutionize how we do things; many others think it's going to take over. Neither side is right, and both sides are. FRT has great power and many applications. It also has deeply embedded biases and represents significant incursions on our privacy. As Vinsel has argued, when we invest too much power in the technology itself we divorce the product from its context. This works for both critique and acclaim: most things are not as bad, or as good, as we think. They are as bad or as good as we make them, which means always remembering the role of "we" humans in conjunction with the technological "it."

For both MT and FRT, there was a clear shift in approach to take people-processing out of the equation. It was never removed entirely: it couldn't be, or the systems would descend into gibberish and ultimately fail. People will always be needed to ensure an approximation of accuracy for data comparison systems, though their role can diminish over time. But that shift from trying to achieve success by reproducing what people do to viewing the challenge as a math problem makes it easier to imagine that people have been elided in the process. They haven't. But their role has shifted and become, in more ways, somewhat obscured; even as all the biases of people remain present in the technology, it takes work to excavate those biases because perhaps we have forgotten the role that people continue to play in the process, even if people-processes are no longer the framework.[50] People and their biases are still there, even as we might have to squint to see them. Biases continue to exist in technologies like FRT as a function of the starting databases from which these machines "learn." Those biases can be minimized once attention is drawn to them, improving the outcomes in an asymptotic curve toward accuracy. The stakes for these biases are the human costs of false positives and people being subjected to false arrest; they make the technology work less well. From a capitalist and surveillance perspective, there is great value in minimizing these problems to get better results: to better identify and surveil people with fewer errors.

That sounds pretty terrifying—like the kind of crit-hype that Vinsel has warned against, partly because it indeed does elide the role of people in the process. But for this particular thing, the tenor is not hype. Surveillance *is* the applied role of face recognition technology. It may be applied to identify people committing crimes, or to exonerate the falsely accused, or to speed up border crossings, or to verify identity in banking to unlock your phone, or for all sorts of reasons that people might support.[51] But all those activities necessarily require an incursion of privacy and the sacrifice of some level of personal liberty. Maybe that's a worthwhile trade-off: many believe it to be the case, including governmental entities and corporations and some portion of the people they serve. Others, as evidenced by the regulation of face recognition in parts of Europe and the banning of governmental and police use of the technology in some US cities, including Oakland, San Francisco, Boston, and Somerville, do not. For some, it's a question of personal liberty in all cases; for others, the line between the Stasi and current and possible future regimes is simply not that dark. Some problems are inherent in the technology itself, including known and unknown biases; others are with the people who apply it, or who could. Some people may be comfortable with commercial uses of this tech while resisting its deployment by national forces. Americans in particular resist governmental CCTV but many have no problem installing ring doorbells that, with slightly different goals, do exactly the same thing.

As it happens, the hype around super recognizer capabilities contains fewer dystopian elements than one might expect, perhaps because the technology at stake here is people. Despite examples to the contrary (ableism, witch hunts, genocides, for instance), it is hard to mount a backlash against something seemingly congenital. But perhaps there is a sense of inevitability about super recognizers: even more than technological infrastructures, people will be people. It's hard to imagine what it would look like to halt the development of . . . people recognizing other people. (I'm not naive here: I appreciate that the use of super recognizers in broad-scale detection is the real issue. No one is trying to stop people from recognizing their older sister's teacher from twenty years ago.)

And yet, we are, just a bit, in backlash with super recognizers now, in the most academic of ways. That doesn't mean the notion of a spectrum has been repudiated; spectrums are a deeply sticky way of thinking about diagnostic categories. While they are supposed to indicate fluidity as per gender and sexuality theory, spectrums also instantiate a binary, with a lot of space and deep divisions between the poles.[52] It doesn't even mean the notion of super recognizers has been repudiated; there is a great deal of research energy and money going into recruiting more of them and demonstrating how, in fact, super recognition works. What it does mean is that the real-world applications of super recognition are being questioned—not on ethical grounds, though that is certainly a relevant and ongoing conversation— but on practical ones. Can they really reproduce in the world what they seem to be able to do in a lab? Or, since so many folks come across the term through internet quizzes—online?[53]

This backlash is related to but not quite emblematic of Vinsel's cautionary note. Face recognition software has serious limitations and shortcomings, and super recognizers lack real-world testing experience and evidence of success, as Sarah Bate and her colleagues continue to remind us in their ongoing super recognizer (SR) research.[54] Their repeated cautions against building a global super recognizer surveillance system is the next stage in SR research, pulling back on the original claims with more tempered applied findings. It's predictable: a new way of understanding something brings first a wave of heated excitement and frenzied research (recall when face blindness was truly invented in the early 2000s), following by some pop culture penetration and crit-hype of the positive sort, and then, another round of crit-hype of the dystopian sort. This isn't a strict timeline; it's messy and a lot of these stages overlap. The nature of this trajectory, which feels practiced and rehearsed, reminds us that even novelty has precedence. Even new things follow well-trodden paths. There are features that link the new to the old; in this case, bias, and revolutionary claims, and the evacuation of political stakes and their reinfusion. Which isn't to say we shouldn't be paying close attention: we should, because the stakes are high. We just have to recognize them. And historical recognition is something at which those in power have not ever been particularly super.

Conclusion: Computers Are Not People and Process Is Not Product

The story of face recognition software and the story of machine translation are the stories, in a way, of misconstruing product as process. Both had their development roots in attempts to make machines mimic people, with the anticipated addition of learning just a bit more about how those human cognitive processes worked. But that approach itself did not work very well. Instead, developers shifted to focusing only on outcomes, turning the goal into the kind of problem that machines handle well, equipped as they are to algorithmically compare data at scale and volume in short amounts of time. That's what both MT and FRT do, and they work as well as they do only because there are real people offering the starting data sets and checking on the accuracy of the comparisons. Terms like artificial intelligence and machine learning have proven sticky, but they are misnomers that continue to encourage slippage or blurring of lines between what people do and what computers do. Just because the final result looks the same doesn't mean that they got there the same way. Even as the category is drawn from what people do, trying to mimic brain processes is inefficient, clunky, and challenging. It's also in some ways impossible because people have choices in how they apply their skills and reasoning. They can, and do, make these choices in the technologies we design, and we can excavate these choices in order to appreciate exactly what these technologies are for. This matters not just because it's good to understand how technologies work, but because as we grapple with the origins and meaning of these technologies and their challenges, it's important to understand that people continue to play a role, even if it's not quite the one we may have thought. And it's important to understand how the origins and goals are a fundamental part of what the technology does, even if, as in the case of MT, it can be applied in a variety of ways.[55]

Human expertise is deeply embedded within the machines, and we must always understand machine output in that context. In a way, the plea to understand what the technology can and cannot do is a plea for the health humanities: in the end, we still have to listen to the experiences of people and put them in historical and experiential context. We have to understand medicine and health care as part of what it means to be human, and we have to consider relationships between people and between people and the world around them as part of what we need to learn about how they navigate space and time. The question is not: "Does FRT work better than super recognizers?" but rather "What is embedded in these technologies?" which is another way of saying "What were they always already for?"

Given the persistence of FRT and its advantages in comparing live faces to stored data at speed and scale, we must be attentive to the ethical implications and actual challenges posed by the technology and those who wield it. Bearing in mind Lee Vinsel's cautionary note that crit-hype can exaggerate threats as much as it can oversell advantages, a historical perspective shows not just what technology can and cannot do, but how to understand the enduring role that people play in how it is applied. Big tech doesn't want these headaches, or wants to head them off at the pass, following public sentiment to apply limited moratoria on some kinds of research and appointing ethics officers to oversee others. It's a cynical move in its way, but also one that is as good as the people in these roles and the power and leverage that they have. That's how capitalism works. And while there may not be ethical consumption in capitalism (or in an environmental crisis), there is always better and worse. As we move toward more systematized ethical audits in computer science and consideration for technological application, we remember that as recently as 2003 and likely even later, these kinds of programming challenges could feel like a fun problem to be solved as if in a vacuum.

But nothing is a vacuum. Every solution brings about new problems. Every answer raises new questions. And let us be very careful who is asking them, and let us remember the presence of people where we see only machines. Let us unblack the box and be sure to recognize, as Os Keyes argued, that we can't separate the technology from the hardware, which we can't separate from infrastructure, all of which are deeply flawed. Even as accuracy improves, these structures still remain mechanisms of surveillance, which is itself tied up in a problematic and fundamentally biased infrastructure.[56] Just as we cannot divide the components from the system, we cannot ignore the people within the components. It is not just our privilege but our responsibility to know exactly what the roles of these people are, and what the machines are, and how they work together.

Not all deployments of biometrics are equal, though they may be equally insidious. These forms of surveillance are a tool of both government and capitalism; tracking people has applications in multiple realms. Some may trust governments more than businesses, while for others, bearing in mind the histories of oppression and discrimination, it is precisely the opposite. And there are many who, when it comes to their own biometric data, trust no one.

Even if the algorithmic biases are reduced, the tools themselves can always be misused. But that is the case with just about everything ever. The tools are also, municipal and EU guidelines notwithstanding, not going anywhere. What remains now—and the landscape is changing even as I write—is to develop and implement a robust and dynamic set of legal guidelines to shape how this evolving technology is wielded and regulated. These frameworks will continue to change as the technology itself changes, as part of the starting algorithm determining the process of machine learning and data comparison, and in conjunction with the human verification and refinement process. As much as we cannot now reject these tools, we also should not imagine that they operate independently. People remain very much a part of these processes that we must unblackbox and demystify, heeding the Algorithmic Justice League's call for "meaningful transparency."[57] The biases in FRT are embedded not just in their starting data sets nor their designers, but the broader structures in which they operate and the surveillance context from which they emerged and continue to operate. By turning to the story of the misnamed machine translation, itself deeply embedded in the Cold War framework in which Bochmann and other face recognition practitioners developed their systems, we can gain insight into when and how the role of people changed in these systems, even as it necessarily has remained central to their goals.

Systems are complex entities whose very existence can sometimes be difficult to identify and define. In chapter 7 we dig deeper to understand how face recognition is a system, and how the working of that system made itself invisible until its failures became manifest. We look at what is obscured by tacit knowledge and what the losses are in not knowing in language and diagnostic structures what we know in experience and relationships. We will explore what is mobilized by the development of a diagnosis, asking if face blindness could even be said to exist without faces, and how face blindness, alongside dyslexia and color blindness, opens up new ways of understanding what was once known and is no longer, and what one day will be known that is not now.

7

Is There Dyslexia without Reading?

Can there be dyslexia without reading? Is there face blindness without a variety of faces? Is super recognition identifiable without cameras? Without the mass production of colored textiles, does color blindness exist? If you never speak, can you have a stutter? Here, we turn to a more theoretical discussion to frame face recognition as something that was itself unrecognized and unknowable, looking at the stakes for its entry as a definable category, and thinking about why that entry took so long. I put face blindness and super recognition in conversation with other categories to unpack how bodily systems work in conjunction with the world around them, centering mediation and crises as necessary components for identification and entry into the documentary record. People's experience of their own bodies is a fundamental part of medicine, diagnosis, living with oneself and others, and being a person in the world. But experience and tacit knowledge can be hard to name without language, and language can be hard to develop without recognition that a thing exists.

These are thought experiments about situationally latent potentialities or perhaps dispositions. We can't ever definitively answer the questions they pose. But that doesn't mean that they don't matter. At the heart of this inquiry is a proposition that certain somatic or neurological conditions are fundamentally unidentifiable, unrecognizable, invisible, and thus cannot be made manifest in absence of some broader interactions with technology, media, and the built environment. Of course, in many ways this is true of all diagnostic categories: as numerous scholars have shown, diagnosis and disease are always functions of broader cultural, social, and contextual interactions with the body and with medical infrastructure.[1]

I've been collecting examples of such categories. In this chapter we consider dyslexia without reading, face blindness without many faces, super recognition without cameras, color blindness without mass production of textiles, and stuttering without speech. Intolerances to food and drink, allergies, and other physiological reactions might also be applicable, but I put them in a different category for a few reasons: first, some are indeed both environmental and evolutionary; their connection to context is grounded, empirical, and in some cases quantifiable. We have a before and after for some examples: before, say, a lactose-free community in Asia consumed animal milk, and after, and what happens; or before and after alcohol was introduced to Indigenous communities in North America, and after.[2] These are important histories grounded in colonialism, racism, medical manipulation, and geopolitical warfare, and they provide models for how to tell stories that include military, geographic, cultural, and physiological history. For the health humanities, there is much to be gained from these approaches, and I urge more scholars to pursue these kinds of narratives. But they are less thought experiments and more key historical insights. Here, I'm after something related but different.

I ask a theoretical question that is narrowly about face blindness and super recognition but broadly about the body and the mind and its relationship to the world: What kinds of categories were once relevant and identifiable and important but are no longer because the built environment and its associated technologies and their associated cultural and social contexts changed? This is slightly different, but related, to Ian Hacking's example of "fugue states," which existed in a particular ecological niche that allowed for their production and identification only in those specific conditions.[3] What will one day be possible and important that is now unimaginable? There are forms of bodily knowledge and bodily interaction that are so deeply contingent they can be rendered invisible and indeed cease to exist (or be called into being) only to once again one day disappear.

This is true of all diagnoses and bodily encounters, most scholars of the body and medicine would say. This position is perhaps the starting assumption of entire fields of study. And my question, my category of analysis, may at first seem to undermine it in its insistence that something like dyslexia or face blindness is a stable neurological entity even absent the conditions of its expression. Why divorce the body from its context? What have we gained if we claim that whatever happens in the brain that makes it impossible for someone to recognize faces is there even if the faces are not?[4] It matters, as we saw in chapters 1 and 2, for the possibilities it opens up about what was once known and is not now.

A Question Is Not a Field

I remember the first time I heard cultural historian Robert Darnton speak. I was in awe. Here was a person who invented a whole field! An entire way of doing history! Would that ever happen again? I wondered. (Yes, it happened again and again.) But we can ask: Did Darnton invent the New Cultural History?[5] Wasn't it happening all along, just waiting for someone to write about it? Of course, there is a difference between the academic discipline of cultural history and the history itself. Or, let us say that doing cultural history is the practice of recording something that was always there, linked with an analysis that was newly applied, the combination of which created a fresh intervention, and a fresh way to intervene. Scholars may invent or innovate fields, but their objects of study had to have existed already. There could be no doing cultural history without culture. And history.

I thought I had invented a bit of a category myself. But no—at least two people have already thought of it: Annemarie Jutel asked the exact question about dyslexia and reading in her book on diagnosis with the provocation that "dyslexia, as one example, may be a disorder in the Western world but would not be problematic in a nonliterate society." For Jutel, the diagnostic category matters less than its implications, as "what a particular group perceives to be problematic or unacceptable, needing remedy, is socially contingent."[6] And Philip Kirby echoed this idea in 2018 in a *History Today* piece, noting that "before the advent of literate societies, dyslexia could not exist in the way we understand it today."[7] But Jutel didn't happen to answer the question or put dyslexia in conversation with other related or similar phenomena. And Kirby hedged with that useful (and appropriate) contextual phrase "in the way we understand it today." I also may not answer the question, but in asking it alongside other examples, I hope to describe a particular kind of interaction between body and environment, between body and technology, between body and media, thinking body *as* media. It's not new, exactly, but when put together in this way, it is newly defined and newly understood. Which is to say: everything in the category already existed, but I put them together in a way that is new and tells us something new about all of them. They haven't quite been linked together this way before. My category itself is about the interaction of a thing and its environment and how that makes something visible that was perhaps always there, but was, prior to this interaction, absolutely undetectable. Face blindness, like all diagnostic categories, underwent a process of invention and contextualization, as we have seen. This particular diagnostic category, so imbricated in mediation, puts an enormous amount of pressure on context for its emergence. It is deeply embedded in a system (face recognition) whose workings are invisible until the system fails.

Sight is a medium and a sense, but it is also a metaphor, and in this case the most appropriate one.[8] So many of these categories engage with sight and its lack: face blindness; super recognition; color blindness; even dyslexia, which was once known as word blindness.[9] That's not, I argue, an accident: all these specific examples are contingent on media for their visibility and identification. This is unlike, say, the whole arena of reactive or allergic or intolerant interactions (to lactose, to alcohol, to tree nuts and legumes) which are about consumption rather than mediation. And in the case of those allergic reactions, I think we can say that the allergy exists even without the nut but perhaps does not exist without the broader environmental conditions that bring it about, and the broader social interactions that bring about the cultural and environmental conditions. But with those conditions in place, the person with the nut allergy has a nut allergy even if she does not have access to nuts. Although, if there are no nuts at all, maybe this is not the case. Except that nuts have been around and accessible for a very long time.[10] Reading has only recently been widely accessed.[11]

"Was there dyslexia before there was reading?" is a different question than, say, "Were there children before there was childhood?" The latter is a classic example of what Ian Hacking explored in *The Social Construction of What?*, an "interactive kind" that he argued is changed or even established by its classification. This does not, he was careful to emphasize, make it less real. But he did make a clear distinction between, for example, a condition like youth homelessness and the people in question, namely homeless youths.[12] Homeless youths exist regardless of the infrastructure around the classification, but the classification and its infrastructure bring a great deal to bear upon the people it enrolls. Hacking also explained the broader contexts necessary for these categories to emerge. He offered "fugue states" as an example of a specific, highly contingent diagnosis that was limited to a particular time and set of places. This ecological niche, in Hacking's terms, provided the conditions and interactions that temporarily allowed fugue states to thrive as a category, a condition, and a lived experience.[13] Naming has "looping effects" that do something to the people in question; in another example, Hacking discussed anorexia as a category that is motivated by fatness and thinness and has effects upon the ones diagnosed with it.[14] The naming and classification itself changes the people, even as the categories and the people are part of a larger set of structures that bring new interactions into being.

Joan Jacobs Brumberg also dealt with the question of eating disorders and social construction, pointing out that bulimia requires access to abundant food and thus, at least in the nineteenth century, would have been limited to the middle and upper classes.[15] So too with anorexia, in the sense that one can be starving and not be anorexic; it is the deliberate avoidance of food rather than the imposed lack of it. Again, this does not mean that the condition does not exist, but rather that there is a meaningful impact of the category of anorexia specifically as opposed to starvation generally, and the former brings with it a host of conditions that change the people associated with it in meaningful ways.

Part of the implications of these looping effects is the notion of the human subject as a category. Numerous scholars have discussed the emergence of this experimental framework that centered on people, offering important dimensions to their somatic experiences that connect to a broader medical, scientific, governmental, and social infrastructure.[16] Conrad and Barker emphasized these interactions in an explicitly medical context, underscoring that reality is not waiting to be discovered but is rather created by those who exist within it, in conjunction with their experiences and environment.[17] They explored the ways that a social constructivist perspective underscores all medical diagnosis and infrastructure, challenging traditional deterministic approaches to medicine.

Charles Rosenberg and Janet Golden have articulated this notion in great depth in their edited volume *Framing Disease: Studies in Cultural History*, which builds upon Rosenberg's foundational work on the framing of disease.[18] This work offers language that contextualizes bodily experience and disease while still emphasizing biological constants. Framing pushes against a notion of arbitrariness, insisting that social and cultural construction cannot be disaggregated from biology. We can also turn to the sociological notion of medicalization, in which a given behavior or way of being in the world becomes classified or pathologized. Foucault would cite homosexuality as a key example; Zola, Conrad, and Szasz offered numerous others.[19] According to these theorists, once the behavior or condition enters the medical infrastructure, deviance falls under the realm of medicine and science, vacating context and infusing medical social control into daily life. Ehrenreich and English also drew on these frameworks to explore the repressive medicalization of women's bodies, particularly around menstruation and reproduction, leading to hysterectomies and other forms of bodily intervention.[20] The notion of medicalization is also tied to Parsons's psychoanalytic theory of the sick role. To Parsons, the position of patient brings with it advantages and obligations that can lead to the cessation of work and other forms of societal responsibility. Parsons approached the sick role as a position of deviance and sought to police those inhabiting it. Other adoptions of the term, and later critiques, have been less concerned with eradicating the role and more invested in understanding the patterns and expectations that accompany patienthood. Anna Cheshire and others have shown how the sick role may impact both the behavior of the person in the role and the relationship of individuals to broader medical infrastructure and systems.[21]

Disability studies has challenged a good deal of the pathological approaches to disease, diagnosis, illness, and impairment. The field has theorized and challenged the medical model of disability, in which disability is rooted in an individual body and results in multiple functional disadvantages that may be alleviated by medical, surgical, and pharmaceutical intervention. The social model of disability, which emerged out of the work of disability activists and scholars, distinguishes between physical impairment and social and infrastructural disability, arguing that disability emerges out of the social barriers that function to make impairments disabling in both the built world and people's attitudes and behaviors.[22] Disability, then, is a product of the environment and can be alleviated by environmental and relationship changes and accommodations.[23] These interventions were extremely important from both a scholarly and an advocacy perspective, giving voice to disabled people to narrate their own experiences, needs, and visions for structural change.[24] Some recent approaches like Alison Kafer's "political/relational model of disability" have explored the interactions between body, environment, and culture, moving away from a strict social model toward a more hybrid approach that still fundamentally insists on rooting disability narratives in disabled experience and perspective. Kafer has troubled the strict distinction between impairment and disability, arguing for a greater attentiveness to chronic pain and mental illness as conditions that are both difficult and disabling.[25] Face blindness is certainly an impairment for which many adaptations exist: the social groups people construct; their modes of interaction; their types of sociality; their self-disclosure; and, indeed, their attentiveness to and ability to understand other identificatory cues. While in some ways these adaptations overlap with super recognition, it's still really hard to be face blind.

But What Can We Now See?

A few much-vaunted and oft-repeated possible cases of face blindness date back to antiquity. I'll again caution here against retroactive diagnoses, in consonance with the smart literature on the topic, while opening up space for the possibilities: we can certainly describe symptoms and experiences that seem to be common, like the struggle or failure to recognize faces.[26] I'll also invite us to consider what we may gain by taking contemporary categories and contemplating them, pushing them, trying to take seriously certain facts of the body that may interact differently with the world in different times and places. We cannot say that this is a manifestation of the modern diagnosis of prosopagnosia, which as we've seen is a collection of highly contingent and contextual factors. Face recognition is a particularly complicated example because there are numerous ways to be bad at recognizing faces beyond what we now call face blindness.[27] But, given that we now think face blindness affects between 1 and 2 percent of the population to varying degrees, the relative lack of discussion of the condition seems puzzling. There are reasons for this lack, though: as many face-blind people have discussed in detail, their condition is one of those things that you don't know is a condition, so it can be hard to identify as even existing. Face-blind people often articulate the enormous sense of relief they feel when their experiences are given a name: they are not stupid, they are not lazy, and they are not rude.[28] But in the absence of widespread discussion of something that is so tacit it is rarely named, they are also not aware that they have a condition. Until they are.

Face blindness seems like a counterexample to the idea that systems remain invisible until they break. People with face blindness manifest a kind of brokenness in a system of recognition and the building of human relationships that in others operates well and smoothly, and in others is visible if unnamed. But in fact, it's an extreme example of the same thing—this system and others like it are so profoundly invisible and tacit that even when they don't work, it's hard to know that they in fact could work and, for others, do. I theorize that this extremity is characteristic of my claims: it takes more than so-called brokenness to render these systems visible and identifiable. If it has always been, for someone, "broken," and for everyone else invisible, then it is hard for the system to be named or known, as indeed was the case for face blindness. Experiential narratives that communicate difference (I can't do this thing and you can) are part of the process, but they aren't enough. Moments of before and after, as in the case of people who developed prosopagnosia from injury or illness, are part of the process, but those moments are not enough, either. A growing need for the workings of the system—in this case interaction with more people and more faces—is part of the process, but it is not enough to enter it into the record.

Together, these things are almost enough. Still, there are some for whom the system does not work. And it is there that the gap lies, and it is there that there have likely been people for whom systems work differently, and about whom we have never known.

I am using the term "broken" in this context to borrow from the theoretical framework, but it is one of my arguments that such system anomalies sometimes indicate a system that is not working, and sometimes indicate a system that works in different and even better ways. There is face blindness, but there is also super recognition. Each is equally invisible and unidentifiable, each requires a form of communicative mediation that works differently from the norm, and each is a system anomaly. One (face blindness) would perhaps be designated as broken. One (super recognition) would not. My framework describes and encompasses both, and perhaps even argues that they are as similar as they are different.[29] Both are invisible system anomalies. Each is valanced differently.

As we saw in chapter 1, Quaglino, Wigan, Hughlings Jackson, and Charcot offer us descriptions of what seems like face blindness, alongside the first use of the term "face blindness" in a medical journal in 1899.[30] It took millennia, and some of that has to do with what studies of the mind and body were interested in, and what was recorded. It is also possible that the condition simply wasn't relevant, because, besides kings and generals who interacted with a great many people on a personal level, there weren't many people to see and know. (Though of course, kings and generals and religious figures were the only people being written about. And who could even write besides them?[31])

So we're at the end of the nineteenth century and we have a few more examples, and then in 1947 the thing gets clinically described by Joachim Bodamer.[32] And Bodamer's case is someone who has the acquired version of face blindness, so that person once recognized faces but no longer does. It takes another twenty years for a further example, and then yet another twenty, and then . . . it becomes something to think of as a condition. It is of course likely that many people experienced the phenomenon described by this condition, but they did not have the language to express it as a difference, even as they lived its implications. It's easy enough—and it's not wrong—to note that the publicity received by Oliver Sacks's powerful writings describing his personal experiences as a face-blind man, as well as the concerted media strategy pursued by researchers of face blindness to attract research subjects, meant people could finally name it.[33] And the internet helped too.[34]

But it wasn't just about the technology and the publicity. It was all these factors together that made this phenomenon knowable and describable, even as it had long been relevant.

Is the same long history also true of dyslexia? Tom Campbell has traced the interest by physicians in this category of condition (or disease, or diagnosis, or learning difference) as linked to a shift in how bodies are judged in the late nineteenth century.[35] As Philip Kirby has shown, the diagnostic category "word blindness" emerged in the late nineteenth century, but diagnosis, research, and services were galvanized in Britain in the middle of the twentieth century with the establishment of institutions and frameworks devoted to dyslexia. Spurred by the Education Act of 1944, which required schooling through the age of 15, Britons were spending more years in school. The "eleven plus" examinations, akin to high school entrance exams, gave select students access to more elite state-funded secondary schools, placing further pressure on educational achievement and the corresponding professional opportunities. Once globally dominant, British manufacturing declined through the 1950s, but the professional classes expanded, raising the stakes for literacy and academic success. Reading began to matter much more. The inability to read—for whatever reason, including language barriers, dyslexia, lack of access to texts or skill development—began to matter much more. Effective reading was a system that was newly near-universal and deeply intertwined with government, business, educational structures, and access. Failure to effectively read was not the thing that exposed that a system of reading existed; there were other explanations for this failure beyond dyslexia.[36] Some subset of people had been reading for a long time, and likely some subset of people had been struggling with reading for just as long. But only in conjunction with a great number of people reading, and the ability to read being required to function, and that ability being something that people had to describe not being able to do, did reading ability itself became visible.

What about super recognition? What about claims that dyslexia allows people to access other ways of knowing and being? (Hendren and Garland-Thomson and Linton and Wendell would argue that with disability, it is always ever thus.[37] If only we make space for other ways of knowing through disability.[38] If only we listen to disabled people. If only we see those who inhabit and know the world in other ways.)

It's not an accident that the word "blindness" comes up so much. These latencies are intertwined with mediation; with the body as medium; with the mediated relationship between the body and technologies of communication; with the systems that encompass all these pieces. "Word blindness" requires words; it requires sight; it requires a system that makes sense of the words through sight; it requires a breakdown of the systems decoding the words by sight. "Face blindness" requires faces; it requires sight; it requires a system that makes sense of the faces through sight; it requires a breakdown of the systems decoding the faces by sight.

We know that media both produce and convey messages, and that systems help. Do bodies make the system of books and reading? Do books make reading? Do bodies make reading? Bodies and books make reading: that's the system. The book is the medium, but so is the body. These are about the interactions of two media to make meaning, and what we are interested in is interruptions to those interactions that make meaning in different ways. Dyslexia can't exist without reading, but the body as medium still exists, and so does a brain that mediates the world. The latencies that get called into being in this way are always mediated, because they are a part of the system. It doesn't matter if dyslexia was in the brain before or without or after reading. What matters is how we understand it. Will it be expressed another way?[39]

But people made words. People did not make faces. These are not the same. Maybe they are the same enough. They both require a system breakdown. Once there was a time when there was no reading. Once there was a time when reading mattered less. There was a never a time with no faces, although there was a time when faces mattered less. There were fewer of them, empirically, and in any given person's life. Only people who saw a great number of people needed to be able to recognize them. We can only posit that for most people, struggling to recognize faces didn't matter much. And, some might say, in the twenty-first-century world in which many people are masked because of a pandemic, everyone is somewhat face blind. In the world of Zoom, faces all come with name tags. Face recognition is far more temporary and contingent than we ever imagined.

What else might one day cease to be called into being?

Creating Categories, or Redescribing Them

These theoretical approaches to the body, diagnosis, and the ways in which they are embedded in broader social and cultural structures have all informed my theorization of this category, or way of thinking about a category. As I approached this category, I was keenly aware of the long history of scholars and activists fighting for recognition that bodily experience and indeed the body itself only exist in conjunction with the world. A set of questions that seek to disaggregate the body from its context not only challenges these scholarly and practical perspectives but might well undermine the real gains made in disability activism and patient advocacy that pushed (and continue to push) against a medical system that has historically belittled, disregarded, and ignored patient voices. This pattern is particularly acute for people who have been marginalized, including women, disabled people, prisoners, people in poverty, and especially people of color (especially in the United States and the Global North).[40] I am sensitive to both the intellectual and political imperative to insist on prioritizing lived experience and to maintain the messy and embedded relationship between bodies, brains, people, systems, environments, experiences, and structures.

Take that as a starting point for thinking about face blindness and its relationship to the conditions of the surrounding environment. And then, let us still ask: Can there be face blindness without (many) faces? Is this condition, which can so deeply impact how people engage with those around them, entirely contingent on there being people around them who cannot be distinguished? The question necessarily implies that a condition (face blindness) exists in some way independent of the world around it. That is to say that somewhere in the brain there is a condition that can somehow be defined even in a world in which it has no meaning. So to insist on examining the existence of, say, dyslexia absent reading, means that the brain's dyslexia—a condition deeply embedded in broader social structures in specific ways—seems in some way to disaggregate the brain-body from the wider world. It seems to raise the possibility, at least, of a stable brain-body condition that always is, even as it waits for a world to catch up to it and call it into being in a visible and meaningful way. It seems to acknowledge, at least, the possibility that reality, or some form of it, can exist outside of its relationship to the world around it rather than cocreating it.[41]

The body exists. It does things in the world.[42] Particular bodies have specific needs that can absolutely be accommodated by drawing on the unique knowledge of the people who inhabit these bodies, and this knowledge and these needs very much emerge out of the combination of somatic, experiential, cultural, social, and environmental conditions. These bodies do not exist or indeed preexist outside of the world, but they do exist within it, changing and being changed by it.

Social constructivism and medicalization are related to the kinds of examples I explore here, but they are not quite the same thing. Medicalization pathologizes and classifies a given behavior, condition, or way of being in the world, but that given behavior existed prior to the imposition of a disease category or diagnosis. With something like self-starvation (for example), the introduction of "anorexia" and "anorexic" upon a person operated in conjunction with a given set of conditions or behaviors; the category changed (and, Hacking would argue, thus changed the behavior and the person doing it), but the manifestation or expression of starvation remained a common thread.[43] In the cases I raise, the conditions of possibility for the behavior or way of being in the world are absent until the world itself changed. It's not a question only of classification changing something from one thing to another (or making it a thing at all) but something becoming visible and identifiable when it once wasn't. The thing may have changed as a function of its emergence and categorization—indeed, it almost certainly did—but even in its invisible or unidentifiable or unrecognizable state, it was, in some way, there.

But this isn't only about behavior that may need a trigger to be expressed but whose possibility was always present or the imposition of labels constructed in conjunction with political and social contexts, but rather a stable biological, neurological, or somatic situation. But can any aspect of bodily experience or existence be stable in this way? These embedded social perspectives on somaticism seem to ignore the world as is, the brain-body that already is and has already been in the world, even if parts of it remain un-knowable. The question does not disaggregate bodies and brains from the world; it just notes that the world can change, and has changed, and the parts of the brain-body that are knowable and are available to cocreate reality in any given moment can be in flux. Again, this is a thought experiment, and in some ways it is a series of unanswerable questions. But in asking, we can unearth those other forms of brain-body availability and in so doing, ges-ture toward what might also one day become known.

Furthermore, maybe there is a way to answer the question about dyslexia and reading, or faces and face blindness, or at least begin to see what that might look like. Because it is about looking, seeing, and forms of mediation. These things, this group of only-sometimes-knowable-and-visible condi-tions, are all about media. Sight, vision, recognition, reading, speech: it is no accident that these are all functions of communication.

The Metaphors of Sight and the Body as Media

This is a story about bodies and medical systems, and thus it must be a story about communication: its possibilities, its limitations, and its me-diation. The face, as I have argued elsewhere, is a medium: it communi-cates meaning both reflexively (this is who I am) and relationally (this is what I see in you). As an index to the interior, it is imperfect: we cannot see the self on the face, but we can fashion ourselves in ways that have visual meaning to others.[44] That's the way it goes with media and meaning: we can produce it, but we can't control it. Everyone makes meaning in their own way. But we all make meaning from faces and bodies. Some signifiers are rooted in racial hierarchies: skin color, for example, is attached to a lot of powerful assumptions and experiences. Some signifiers are deeply prob-lematic, rooted in structural oppression with long histories; so judgments made about others based on their skin color are racist, and oppressive, and problematic, with consequences for the lives, safety, health, and dignity of especially people of color. Powerful meanings and associations people have to their cultures and histories are also represented (imperfectly) by skin color, racial and ethnic affiliation, and culture and community. Given the legacies and current instantiations of racism, the meanings attached to skin color usually serve to reinforce structural oppression. But at the same time I do not want to evacuate the lived experience of identity affiliations, many of which are mobilized through racial categories, assigned or assumed.

So too with other bodily signifiers. Some have richer and more difficult narratives attached to them than others. Some change dramatically over time and across spaces.[45] Blue eyes may have saved a Jewish person's life under the Nazi regime; lighter eyes carry fewer life-and-death implications right now.[46] Eye color is an imperfect signifier, but it is a powerful one. And one that reveals that perhaps all such signifiers are imperfect: Jews can, of course, have blue eyes.[47] And blue eyes do not make someone a more valuable person. But people can make meaning out of blue eyes that has serious consequences in the world, or at least did at one point. Countless examples span every possible spectrum, many of which feel unpredictable: looking like Kim Kardashian (either naturally or surgically) can make someone a lot of money.[48] Which is to say, the faces and bodies we inhabit matter, in large part because people—including the ones inhabiting these faces and bodies—make meaning from them.

John Durham Peters took up the embeddedness of bodies, beings, environments, and entities as networked in his masterwork *The Marvelous Clouds: Toward a Philosophy of Elemental Media*. As Peters eloquently stated, "Wherever data and world are managed, we find media."[49] Our bodies, he continued, are "technical systems as strange and mysterious as any devices we use."[50] For Peters, technical systems are a particular formulation that he explained in depth, but here it is useful to think with his framing around how we imagine the body's interactions work with everything. Peters situated the transition of media to being about human signals in the nineteenth century, as manifested in particular with the Spiritualist figure of the not-accidentally-named "medium" who was the conduit for the transmission and communication of messages (from the dead to the living.)[51] The twentieth century, Peters charted, became focused on mass media; he calls for an expansive vision of nature-as-media that has, he argued, persisted throughout these transitions.[52] Peters traced nature as media; here, I remind us that the body is nature. The body, then, is both a collection of data and a source of its production that finds meaning in its interaction with the world.[53] Like mass media that both produces and conveys messages—think in particular of journalism, journalists, and journalistic corporations who make the news even as they report the news—so too does the body itself receive, produce, and communicate information.[54] And thus, the body is media.

The body is a system. And, like most infrastructural systems, its workings are invisible . . . until they are not. Peters framed his call to make things visible in consonance with Lisa Parks's notion of infrastructural concealment, and classic Science and Technology Studies (STS) work by Bowker and Star on how systems obscure their own workings.[55] Sybille Krämer argued that communication is a process of partially revealing the hidden: the medium itself helps connect across disparate approaches, serving as the messenger. But communication can only ever be partial, and the messenger can only ever reveal the trace of the whole; it is up to the receiver to respond, even as the receiver acknowledges the limits to complete understanding. Communication, then, is an ethical relationship between the receiver and the other; that other may, as we have seen, be non-human, but communication may well indeed be between two (or more) people with the body acting as medium.[56] The body then, is medium, messenger, and message that always necessarily requires an other; communication with and of the body is meaningful only in the context of a relationship, only in the presence of a receiver. Our bodies are meaningful in communication, as communication, as media, as conveyances to other entities, which can also include ourselves. The system is a fractured one.

Like all media, bodies are imperfect technologies of communication, whose workings are obscure, variant, and manipulable. We can call the meanings derived from the body socially constructed, as indeed all meaning is. But the media of the human body have special status in that they are always and necessarily inhabited by people, living, dead, and otherwise.[57] We play a role in how we fashion and manipulate and infuse these media with signifiers, but we cannot control how they are understood or absorbed.

Bruno Latour's actor-network theory (ANT) considers the ways that the human species has always been, in its terms, "networked," embedded within places, environments, other people, other species, other things.[58] Strong resonances exist between ANT and Deleuze and Guattari's notion of assemblages, which builds on French philosophical theory to explore the intertwined arrangement of concepts and beings, and the meaning that emerges from that arrangement with respect to those specific entities.[59] These arrangements encompass the relationship between human and non-human entities; no one entity, human or otherwise, has a greater ontological status than any other at the outset. These arrangements are temporary and always relational, such that meaning exists in the context of both how these entities are arranged and what that arrangement calls into being. And that arrangement can and will always shift. Numerous scholars operating in a justice and equity framework have built upon these theories: notable are applications by feminist geographers Rachel Colls and Rachel Slocum's queering of these ideas alongside Jasbir Puar's expansion of the notion to explore how race, gender, sexuality (and, later, disability) are themselves encounters and events between entities rather than (unstable) subjective features of given bodies and beings.[60] Assemblages help us think through the meanings that emerge from the interactions of body/world alongside the meanings that are produced for body and for world as part of these interactions. To return to my original thought experiment, we can situate dyslexia as an assemblage, which would be consonant with the commitments of those who underscore the embeddedness of the body in its context along multiple perspectives. What is gained, however, by understanding the body itself as a medium? And what kind of thing, in this formulation, is reading? Or face blindness? Or super recognition? Or color blindness?

And how can we find out?

Peters used to great effect the STS analysis of the way that the workings of many systems and processes are invisible until they break; in his formulation, the earth and all that it contains is the media infrastructure that is now ceasing to function.[61] And in that slow and steady demise, much that was present but unknown can now be seen and known even as it ceases to exist. I ask a related and opposite question: What of the systems and processes and infrastructures that are invisible *because* they are not now working together, but can be recognized if and only if the assemblage has multiple modalities of meaning? This formulation understands diagnoses quite differently; they are not in and of themselves good or bad: their valence is part of the production of meaning only in relation to the other modalities in which they exist. Or, to break that down further with my thought experiment: dyslexia is one assemblage of text and mind-body; another can be traditional reading, and another can be non-reading.

Reading is a system that is visible in success rather than failure; dyslexia is not a distinct system of only failure, but one that is related to text and brain-body and visible with respect to text, to brain-body, to eyes, and, as Tom Campbell and Philip Kirby have both shown, to a coproduced system of capitalist value that places a premium on the translation and communication of written text to knowledge production.[62] Dyslexia, then, is many things: a category that is mobilized for support and services and explanations; an artifact of a changing system of capital and production; more recently, a way to access alternative ways of learning and knowing; and, maybe in Peters's terms, a failure that makes the success explicit.[63] But can it also be something else? Or, more exactly, if we think of dyslexia (and face blindness) as assemblages, what are its component parts? Are they the brain-body and the text, or can we get more granular? Is there something within the brain-body that is unique to dyslexia, in contrast to the components unique to other bodily assemblages?

If we can find a way to talk about those component parts, we can find a way to unlock other invisible systems that were once extant and are no longer, and perhaps think about potential systems that one day will be. And in so doing, we can take seriously the provocation that what medicine deems failure or diagnosis or disease is not necessarily so. Some systems become visible only when they break down, but others become visible only when something is built. This is of course obvious in some ways and is predicted by the territorialization and deterritorialization processes articulated by Deleuze and Guattari around material forms and forms of content.[64] Except that if we explicitly understand the body as a medium within this framework, what can we now know? And does that formulation lend insight into why the particular conditions in question are all themselves media assemblages that are explicitly relational and communicative? John Durham Peters would perhaps say no: his argument posited an expansive understanding of media that would take the question as a given. But I want to push further and see what else, if we know this, we might be able to know.

Conclusion: Seeing Like a System

These latencies are not all diseases, and they are not just diseases or disabilities. If we think of them together, putting super recognition and face blindness and dyslexia and color blindness in the same category, we can see the ways that they can help people see that which is to others obscured. This is not the same as narratives that situate, say, mental illness as a key to creative genius or even to the advantages that can come from a non-neurotypical way of engaging the world.[65] Philip Kirby has chronicled cultural representations of dyslexia as a kind of gift, a way of knowing that only a limited few can access. He cited in particular the Percy Jackson young adult novels, in which the protagonist's dyslexia is both a godly inheritance and a tool that helps Jackson in fulfilling his destiny. The author, Rick Riordan, began the series in part to offer his dyslexic son a different way of understanding his learning difference.[66] It's a powerful adjustment, and also a fictional one: while many kids struggling to read will see themselves in the protagonist's scholastic experiences, few of them are also gods.

The neuroscience research often locates varies compensations and adjustments in non-neurotypical expressions: people with various synesthesias may report, for example, deficits in other arenas.[67] As we've seen in chapter 4, in order to identify such deficits or compensations, researchers have to be interested in looking for them, and that interest is indexed to whether they consider a certain condition to be a lack or a benefit; there has been little investigation in the challenges that might accompany super recognition, for example, and there has been correspondingly little interest in the adaptations and compensations that face-blind people may exhibit.

Face blindness and dyslexia are not brokenness, and super recognition is not hyper fixedness, but they are ways of accessing the world and its media that work differently. But the world, as we know, changes. What once might have been a way of working that didn't work may soon be the best way to work. When the world changes, a system whose operations were inefficient in context might become more effective as contexts change. And then the system itself must, through this reevaluation and its looping effects, change. Rethinking these mediated differences with the body as media opens the possibility that these knowledges might in other contexts, with other receivers around other assemblages, be the better or at least more efficient or maybe even more invisible system.

These systems that we rely on and operate within as though they were the best possible systems are not in fact the best. We know this. We know structural racism and sexism and ableism are themselves not a bug but a feature of these very systems. But it may take someone for whom the system's failures have been personal and manifest to imagine otherwise. I think here of Octavia Butler's stunning Xenogenesis trilogy, which, among other innovations, finds in cancer a talent rather than a disease. In her vision of the world to be, cancer offers the potential for cell regeneration and growth, a capability that is of enormous value to species looking to cure injury, suture illness, and occupy new ways of being in the universe.[68] The protagonist is specifically selected for this talent that she feared would be her death sentence—and it may well have been, in her context. The novels change the context. Rapid cell growth becomes salvation. Cancer becomes salvation.

But the cancer was already there. The systems changed.

The discourses around the social construction of disease, medicalization, the sick role, and the different models of disability all draw on the meanings that bodies communicate and the various pressures and powers that come to bear on how we make sense of and derive those meanings. They all rightly insist that disease, illness, conditions, disability, and diagnoses are negotiations between these perceived meanings, medical power structures, social, cultural, and political developments, and historical assumptions and traditions. There is a way to make these claims that evacuate the body or render its corporeal form unnecessary. There is also a way to make these claims that has sympathy with lived experience and takes it into account.[69] There are tensions here: Riva Lehrer wrote movingly about feeling liberated to discuss her pain within the Disability community, challenging a tradition that was concerned that acknowledgment of pain would undermine the argument that Disability is not inherently a disadvantage and is only rendered thus by the lack of social and environmental accommodation.[70] Two things can be true: pain is terrible, and Disability need not be fundamentally disabling.[71]

Is there a meaningful difference between visible parts of the body—the number of limbs, eye color, various measurements—and ones that cannot be seen or felt without medical intervention or particular experiences, like allergies, intolerances, heart conditions? What about temporary or transient conditions that mark themselves upon the body, like hives or rashes that may or may not be attached to specific diseases like chicken pox or eczema? Are these conditions equally the results of social construction and broader meaning-making? Perhaps eye color is invisible or unidentifiable or meaningless under certain conditions, and thus cannot be known or experienced, much like certain disease categories. Which is a different way of saying that they exist only in context, like all somatic categories, diagnostic, pathological, or otherwise. Which is a different way of saying: this thought experiment is not about good or bad. Furthermore, our categories of disability and ability are themselves deeply contingent and highly mediated.

A thought experiment is only as nuanced as the meaning infused in the categories themselves. Comparing neurotypical or non-disabled ways of interacting with the world with neurodivergent or disabled ways of being often takes a deeply reductive position as its starting point, failing to account for the rich and highly varied lived experience of disability. I take seriously the complex interactions between personal experience, medical and diagnostic infrastructure, and history itself as an actor in the mediation and categorization of disability. When we frame conditions such as dyslexia and face blindness as historically mediated and experienced phenomena, we open greater possibility of what was once known and is no longer, and, importantly, what can one day be known that is not now.

The category of dyslexia tends to be indexed primarily to text and reading; if we frame it as a complex mediated category that includes not just the processing but the production of knowledge, we have a great framework to understand, for example, the role it may play in auditory processing, or spatial arrangements, with implications for navigating space.[72] We can turn to first-person accounts of face-blind people or dyslexic people to learn about both the experience of navigating the world with modes of compensation and accommodation *and* what and how people know. This chapter is both a thought experiment and a call for an expanded epistemology informed by an understanding of mediated neurodivergence as a category of knowledge production itself. In a non-literate culture or in a context of few face-to-face interactions, the implications of disability and face blindness will manifest differently, and likely not as disabilities. This speaks to both the social contingency of disability and our own limitations about thinking broadly about ways of knowing and ways of being.

What can face-blind people know? What can they see? This is a function of what there is to see, of course, and how it is seen. The brain-body works in conjunction with the structures through mediation to produce specific forms of knowledge. That is only meaningful if that knowledge can be communicated, and it is only valuable if that knowledge has value. In this case, face-blind people tend to be enormously skilled at reading body language, gait, facial expressions, and voices. They tend to have highly diverse groups of friends and prioritize unusual appearance; these are adaptations, but they are also important skills and, one could argue, ethical frameworks.[73] Some might say the same is true of color blindness. Once those who were color-blind were actually able to see minute variations that helped enormously with hunting.[74] That ability may one day again help with something else. We don't know yet, but that's not the same as it not being so.

Conclusion

Beyond the Face

What's the lesson here? We have built a history of face recognition; tracked its invention across a spectrum from face blindness to super recognition; offered the biography of two extremes; learned about the lived experience of those on the poles; explored the long history of biometric surveillance; and emphatically insisted on the role that people play in these processes and their invention. We've reflected on the ways that local and individual stories of people like Richard Russell, Josh Davis, and Jennifer Jarett become official histories, and we have been reflexive about the role that those writing and recording those histories (in this case, me) play in the process alongside key researchers and subjects. We've looked at grounded examples of the relationship between surveillance and face recognition at Checkpoint Charlie and considered the ethics of unearthing these stories. We've thought about the relationship between machine translation and face recognition technology in the Cold War context and how that framed applications of these structures. We've traced the role of media in inventing and developing face recognition frameworks, and how key high-profile figures Oliver Sacks and Chuck Close helped build awareness of the category of face blindness itself. We've examined the ways that tacit knowledge remains invisible, and the difficulties in naming that which elides identification and recognition. We've insisted that naming helps. To a point. And we've noted the value of considering extremes together and the ways that spectrum thinking works against that tendency, labeling and valuing the poles in opposite ways. We've also noted the limits of that which links extremeness together; some things are more challenging than others. But even still, as with face blindness, there are adaptations, compensations, and, in the words of Rockstar Dinosaur Pirate Princess, the ability to read faces *like a boss*. For super recognizers there are also difficulties—it's a world of data that they can't turn off—but it is, at least according to Gary Collins and others, utterly worth it. But, super recognizer Jennifer Jarett is quick to remind us, the ability to recognize faces better than other people, largely independent of relationships with them, is not actually a superpower. And actually, as researcher Sarah Bate continues to investigate, that ability has its limits.

There are stakes for face recognition. Not being able to recognize faces is hard, both in a practical day-to-day sense, and also in terms of building relationships, relating to others, developing shared memories and shared history, and, of course, the sheer fact of knowing who other people are. Being online helps. In ways we could not have predicted, Zoom, with its labels appended to specific people, helps. Having friends who look significantly different from one another helps. Face-blind people tend to have significantly more diverse groups of friends, with a wide range of gender expression, racial and ethnic identities, and unusual hair colors, piercings, and other identifiable markers. It makes sense: if you have only one friend with red hair, you are more likely to be able to correctly recognize who she is. Furthermore, face-blind people seem to prioritize traditional notions of beauty much less than other people, partly because so many people try to adhere to these norms, and if people have similar norms of beauty, they are harder to tell apart. If everyone is striving toward straight light hair, a thin body, and roughly symmetrical features, everyone will start to look alike. This is precisely the opposite of what face-blind people want in those people with whom they choose to surround themselves. Of course, we can't always pick the people we love or the people we meet. But it's worth paying attention to this surprising path toward diversity across a variety of categories; it represents an adaptation for face-blind people, but also, perhaps, an adaptation that could be useful to us all.

It's different for the supers. Their tricks and tools and techniques are all about blocking out, or at least pretending to block out, the faces they might inappropriately recognize. For many of them it is a funny sort of skill, a party trick at events and while watching television. Only recently, with the clinical categorization of super recognition converging with a more extensive camera-based surveillance infrastructure, has there been a concrete opportunity to leverage this ability. Some supers are excited to partner with police forces and technology companies to improve the mechanisms designed to identify people largely but not exclusively in the name of security writ large. Others aren't particularly interested, and indeed are quite skeptical of the claims that super recognizers are effective outside clinical settings. But it's all rather new, and the research is ongoing.

While the search to monetize super recognition continues, face recognition technology becomes ever more ubiquitous. The latest in a long line of biometric surveillance mechanisms, it's become much more accurate, with fewer false positives and more definitive matches between the people captured by the tech and the databases that store our faces and identities. The COVID-19 pandemic has changed how we think about exposed faces; the technology, with its ability to focus on and analyze discrete features that remain uncovered even while others are masked, is adjusting. While there are clear improvements in algorithmic bias, the technology continues to concern many of us. The histories of surveillance are themselves deeply biased. As are the hardware infrastructures attached to the software itself. As is the practice of deciding who should be watched; who should be allowed in; who should be a source of concern. And why.

Living on the Edge

As we conclude this story, we should think about a number of points that face recognition and its history help us understand or exemplify. We should note that diagnoses are invented in conjunction with many structures, including medicine, media, cultural context, and key players. We have also learned that while spectrums can be fluid, they also have extremes, and those extremes can be quite sticky. Spectrum thinking encourages placing those binaries in opposition to one another, and such thinking is a loss; face blindness and super recognition bear similarities around forms of sociality and the (lack of a) role of memory as an index to recognition. Extremeness itself is something these two poles have in common, and we should be attentive to that fact. We also need to be attentive to the health humanities more broadly: face recognition took a long time to be invented, partly because it was so tacit. Without a category, face recognition was almost impossible to name. But the experience could be and was described; we just needed new ways to listen. And, of course, once these categories were invented, they were named. And the naming gave people an important tool by which to understand themselves and explain themselves to others. While naming is not in and of itself a form of treatment, we can think of it as a form of therapy, because it can indeed help.

Supers have a huge social advantage once they learn when it might be in their best interest to suppress their abilities. Their skill helps in a variety of professional fields, particularly the service industry and criminal detection. If it constrains where someone can live or shop, it's a fair trade-off. The challenges as we know them to date are dwarfed by the benefits, at least according to the way the condition is presented across multiple media and scholarly literature. The spectrum model places super recognition and face blindness in extreme opposition; the challenges and advantages and experiences are thus opposed as well. One is a superpower and one is a disability, and the binary between those categories is stark. Super recognition is good. Face blindness is bad. Superpowers are good. Disabilities are bad. There are a lot of value judgments laden upon this diagnostic spectrum.

But really, the two poles have a lot in common.

The ones who can never forget a face are on the opposite end of the spectrum from those who can never remember one. But on both sides, the face does not serve as an index to relationships; for both face-blind people and super recognizers, the extent to which they recognize others is a meaningless indicator of their connection to those people. That can change how they relate to others and build relationships. Face-blind people may retreat into spaces where faces and identity are more obviously connected, or they may simply retreat away from others. But they also develop ways of creating instant intimacy and warmth independent of recognition. It's a hedge: no one wants to be not recognized by their best friend; but it's also really nice to always be greeted so warmly. Supers likewise may retreat from the overwhelming amount of information that a sea of people presents, preferring blank spaces and quiet places. But they too may become masters of the immediate connection, not needing historical shared experiences and deep relationships to greet someone with total recognition. It makes sense that supers would gravitate toward positions where face recognition is a huge asset: border guards and cops, of course, but also politicians, fundraisers, and high-end administrators. Prosopagnosiacs may select for the other extreme; anecdotally, a lot of academics are face blind. (Academia's a safe place for slightly socially awkward people who never seem to remember you.) It may seem counterintuitive that there are also face-blind politicians and people in positions of extreme social exposure and responsibility. But it makes sense: at the extreme ends of the spectrum, relationships work a bit differently. Face recognition is so fundamental to how we connect with others; outliers may enact these connections in other ways. We don't know for sure, and I'll leave it to Josh Davis and his colleagues to test it in their way, but I've theorized it in mine.

It's been hard for both the supers and the face blind to recognize that they sit at extreme ends of the face recognition spectrum; people who fall in both categories just assumed others were a bit (or a lot) better or worse at the very basic human ability to recognize others. Getting a diagnosis, in both cases, was a relief, but with different stakes: people with face blindness needed their condition to be named to feel relief; many people with super recognition welcome their diagnoses as an opportunity to capitalize on a skill.

The downsides to super recognition, as presented in the television news interviews and print articles, are small and humanizing ones that these supers can deal with easily and with grace. They are the ones that supers themselves bring up: awkward encounters with barely noticed passers-by, extra care in where one lives, the need to leave a mall quickly. There has been almost no scientific research in the possible social or cognitive downsides of these abilities. Studying these extremes together can offer a valuable way to decenter face hierarchies and challenge their underlying biases, while at the same challenging the binary at the extremes of spectrum thinking. If we pay more attention to what gets mobilized by naming these extremes along a spectrum and categorizing the people within these extremes, we can identify and maybe even decode the associations compounded by the categories of disability and super ability. Face recognition extremes underscore that everyone sees the world and recognizes it a little differently. That's the same across the spectrum. If we know more about how others view, and recognize, and process something as socially powerful and indexically loaded as the face, perhaps we can have a more expansive notion of experience, identity, and relationality. But the deeper implications of recognition in absence of emotional connection, and the possible deficits that may balance the clear cognitive advantages of super recognition remain, at the moment, unstudied and undiscussed.

Beyond the Face

In the course of researching and writing this book, I began to wonder: What would it look like to live in a world without faces, or a world beyond faces, or a world in which the face itself was redefined? We make a lot of meaning from the face. From eye shape to brow elevation; nose size to lip alignment; skin color to hair texture; cheekbones to jawline; wrinkles to taut skin; and on and on, we make subconscious note of facial features and run them through a mill of cultural and personal assumptions. Faces have a special status. I should know: I've spent my whole career as an academic and professor studying the history and science of what faces mean to us. Here's what else I know: faces tell us a lot more about the people looking at them than the people sporting them. The snap judgments made through facial reading have myriad implications for how we classify people, how we evaluate people, and how we value people. We imagine that we know something about the character of others from how they look, how they fashion themselves, how they appear to the world. And we spend a lot of time worrying about our own faces and what they might mean to others.

This book tells the story of face recognition: its history, its cultural impact, its value, its role in human interaction and communication. It's the story of what happens when we use faces to define a category of recognition, and its implications for interaction, technology, and relationships. And it's the story of the limitations of the face: what we can't know from looking at it, what we can't do by changing it, and what we can't learn by studying it. At its heart, it's the story of what it means to listen to our stories about the face.

We learn early on in life that "better-looking" people get more stuff. More attention, better grades, more latitude to make mistakes, more forgiveness, more wealth. And "better-looking" historically has meant faces that look whiter, younger, more "relatable." There is no meritocracy when it comes to the face. And in a world increasingly turning to biometrics for security and surveillance, our facially driven identities take on another dimension of importance, raising innumerable ethical issues including the assumptions baked into the code behind the technology.

What if it could be otherwise? What if we could imagine a world where we made sense of people in ways that were not so intimately tied to our biases around the face? Would we still know others for who they are? Would we still know ourselves?

I can't answer that question, except at the very margins, through the eyes of people who literally can't recognize faces. And I can't know what it would be like if faces were infinitely mutable—something to switch up in the same way we change glasses or jeans—except by exploring the limited cases of face transplants and extreme plastic surgery. I can't speculate what a world of people with the same face would be like. But I do know this: it would be very different than the world we have now. We rely a *lot* on faces. So, to imagine the contours of human interaction that isn't based on faces, we have to consider all the ways in which that interaction is currently subject to faces. Only then can we imagine other possibilities, in which the faces we have need not dictate the narratives of our lives, and the faces we see on others need not frame the people we imagine them to be.

Our ingrained ideas about faces matter: they reproduce racial and gender biases about better ways to look, which, knowingly or otherwise, we often translate as better ways to be. But we don't have to change our faces to conform to who we think we should be, or how we want others to see us, or even—as in the narrative of every makeover television show *ever*—to match the person we actually are on the inside. Because it's possible, and even probable, that our faces do not effectively communicate who we are. Which means that we absolutely need to change how we think about faces. And the stories we tell about them. And who gets to tell those stories, and who decides to hear them.

In the world of *Orphan Black*, a BBC American television show about clones, facial features themselves don't tell us much about who is who. Most of the relevant characters have precisely the same face. So the show works at the level of dialogue and gait and costuming and hair to make sure the audience distinguishes one clone from another. In a weird twist, the show is actually less confusing for face-blind people than for others, because they have always had to pay attention to those kinds of external cues, while everyone else is just learning these skills. That moment of encounter is confusing, of course, because face-blind people don't automatically realize that the tension lies in the fact of facial twinship. So, in a thoroughly modern moment of media consumption, such scenes are spoiled on the online forums devoted to watching *Orphan Black* while face blind. (There truly is something for everyone online.[1])

In the chapter on dyslexia without reading, I asked what forms of knowledge are now obscured because the media infrastructures around them have changed. The *Orphan Black* example is another way of asking this question: What are skills that we cannot recognize because we have already decided what is a "good" and a "bad" end of a spectrum? That is not to minimize the challenges associated with some impairments, but rather to urge us to think more creatively about different ways of interacting with the world.

I'm not talking about blank faces or a world of masks or the convergence of Instagram face so that everyone looks the same online and IRL (in real life). I'm not advocating interventions that make everyone look the same so that the face means nothing. That's far from ideal. And it wouldn't work: in a world of clones, even the smallest distinction matters. People always find ways to be unique, even if unique means looking the most like an (often imagined, largely manipulated, generally impossible) ideal.

So what, exactly, is the world beyond the face? It's not quite faceless: we have faces, we should have faces, we will always have faces. And those faces matter. In ways good and bad. They matter as something that is most intimately ourselves, reflecting our histories, both personal and familial (yes, these are my mother's cheekbones and my grandfather's eyes), our decisions (remember your sunscreen, kids; you'll thank me for it when you are old), and in many important ways, our identities. Skin color matters, and it should, insofar as it is part of everyone's personal and collective story. But skin color also matters in ways that it shouldn't, giving people with the right (light) skin myriad advantages both explicit and implicit, both individual and collective. In a world beyond the face, the face matters only as one part of who we are. That's how we imagine it is now, but, as this book has chronicled, that's far, far from the case. Faces do all kinds of work to categorize and classify (and sell stuff!) and to include but much more often, to exclude. To decide (in life as in fashion) who is in and who is out.

I think there are other ways to make sense of others. I think the world beyond the face is the one where faces are a collection of features. And those features tell a story. And it's a story that we don't imagine we already know. In a world beyond the face, we take the time to ask people: What's your story? Who, actually, are you? Because even though I see you, I don't already know.

Coda

The Detective Story

I couldn't resist: I did, in fact, write a bit of a detective story, tracking down a possible case of face blindness and filling in the prehistory to the history I've offered in this book. But the truth is that it doesn't really belong. It's one anecdotal story. It doesn't contribute either to origins or narratives of invention. It's exclusively Western and based entirely on limited material. But at least I included it only in a coda. And, honestly, I think it shows that even though the tacit nature of face recognition made it hard to identify it as a category both along the spectrum and at its extremes, it was not impossible. This before-and-after case from antiquity highlights both that people could once identify others and lose that ability and that this loss could be named. Which means that it could be, and once was, known. And came to be known again.

Athenian historian and general Thucydides (460–400 BCE), in his recounting of the war between Sparta and Athens in the fifth century BCE, described the plague in devastating and careful detail. He charted geographical spread, symptoms, emotional impact, and long-term consequences. In an acutely resonant comment, Thucydides noted the limits of his own knowledge and capabilities, leaving "individual opinions of the likely origin of the plague, and of the factors which they think significant enough to have had the capacity to cause such a profound change" to "others," including "doctors or" (tongue firmly in cheek) "laymen." Instead, he "shall simply tell it as it happened, and describe the features of the disease." His goal was pedagogical and historical, recording these details "to give anyone who studies them some prior knowledge to enable recognition should it ever strike again." He was well situated to offer these descriptions, having himself "caught the plague, and witnessed others suffering from it."

With this careful introduction, Thucydides continued with his descriptions, discussing who was most likely to be afflicted and how the plague manifested, from the "high fever" and "reddening and inflammation of the eyes" at the outset through "small pustules and ulcers" to, eventually, "heavy ulceration and the onset of completely liquid diarrhoea." For the very few who survived this stage, the disease moved to the extremities, attacking "genitals, fingers, and toes." These effects persisted, "and many lived on with these parts lost." Many "too lost their sight," both literally and in more abstract ways: "There were those who on recovery suffered immediate and total loss of memory, not knowing who they were and," particularly germane for our purposes, were thence "unable to recognize their friends."[1]

From this statement much has been made. It's useful of course, because it is old, and careful, and specific. It described a course of illness that includes a fever, shortness of breath, cough, bleeding, violent spasms, vomiting, and, eventually, for some, loss of limbs, loss of memory, loss of recognition. It was a specific type of memory loss, notable enough for Thucydides to make note of it in addition to the general recall deficit and its implications. What we can say from this description is that some survivors of the plague suffered many long-term consequences, including a general loss of memory and a specific loss of recognition of friends. We do not know if this loss included recognition of people in general, or only those who were already known. What we can reason is that the fever likely had cognitive effects, some of which were quite serious. We don't know how long they lasted or how extreme or wide-ranging they were: we do know that Thucydides highlighted lack of face recognition in particular as something that people once had and then had no longer. Face recognition may or may not have returned, though of course life spans were shorter; many would not have had much time in which improvement could manifest. But still: friends could no longer be recognized. Something was lost. Something meaningful.

This isn't exactly a "case." It's a description of general symptoms and effects. We don't have an individual to follow, to interrogate, to use as a point of comparison. We have "those on the road to recovery" who were "unable to recognize their friends." We don't have a single instance, a particular collection of events, symptoms, experiences, results. If we don't have a case, can we reason or think with it? We do have something. We have a point in time. We know that after a fever-causing plague, survivors continued to suffer various effects including, sometimes, loss of recognition. Of their friends.

So, loss of recognition following illness is a condition for which we have a historical record. That sounds a lot like what happens with acquired prosopagnosia. It can also sound a lot like something else: as we see in chapter 2, there are many ways to be bad at recognizing faces. But this is an acquired deficit following an infection or illness. That we can say for sure. And we can also say that there was meaning in recognizing friends . . . and subsequently not being able to do so.

Thucydides did not say anything else about this condition. That's all the information we have. It's not a lot, and we don't have much more in Western literature for a long time—a *really* long time—right up to the cases discussed in chapter 1.[2] It's possible I missed something in that couple of thousand years, but maybe not: as we have seen, it's been hard for people to identify that they have more than average trouble recognizing faces, and even harder for people to know that face blindness is even a possible category. The inability to recognize that face blindness was a condition would have been so much more pronounced before it was, in fact, officially a thing. It's impossible to reason in cases without . . . cases. But this example is both a good story (and stories matter!) and an example that opens up the possibility of stories and categories that we do not yet know and do not yet know how to identify. It also highlights a thread of continuity around the stakes for recognizing faces; that, we see from Thucydides, has always mattered. Not being able to recognize faces—due to fever, or from birth, or because of other types of impairment or experiences—has always mattered. This ancient example was framed in the context of illness, the helpful before-and-after situated around a crisis that helped identify a category of experience that otherwise remained too tacit to name. As we have seen, there can be power in being what we now call a patient, even as the patienthood is a sticky classification that leverages all kinds of associations.

But this is true for people who became face blind because they became patients. How do we think about those for whom face blindness has always been, and patienthood has not?

Well, as this book shows, we didn't think about them for a long time.

Notes

Introduction. Inventing a Spectrum

1. Sharrona Pearl, *About Faces: Physiognomy in Nineteenth-Century Britain* (Cambridge, MA: Harvard University Press, 2010); Sharrona Pearl, *Face/On: Face Transplants and the Ethics of the Other* (Chicago: University of Chicago Press, 2017).

2. Paul Lucier, "The Professional and the Scientist in Nineteenth-Century America," *Isis* 100, no. 4 (2009): 699–732, https://doi.org/10.1086/652016; Ruth Barton, "'Men of Science': Language, Identity and Professionalization in the Mid-Victorian Scientific Community," *History of Science* 41, no. 1 (March 1, 2003): 73–119, https://doi.org/10.1177/007327530304100103.

3. Here I follow Katherine McKittrick, *Dear Science and Other Stories* (Durham, NC: Duke University Press, 2021).

4. See, for example, Keith Breckenridge, *Biometric State: The Global Politics of Identification and Surveillance in South Africa, 1850 to the Present* (Cambridge: Cambridge University Press, 2014), https://doi.org/10.1017/CBO9781139939546.

5. Saidiya Hartman, "Venus in Two Acts," *Small Axe* 12, no. 2 (2008): 1–14.

6. Seeker, *How Exactly Do Our Brains Recognize Faces?*, 2016, https://www.youtube.com/watch?v=Y1lnrGIbweY.

7. I am grateful to Courtney Thompson for sending me this reference.

8. Heather Sellers, *You Don't Look Like Anyone I Know: A True Story of Family, Face Blindness, and Forgiveness* (New York: Riverhead, 2011).

9. Oliver Sacks, "Face-Blind," *New Yorker*, August 23, 2010, https://www.newyorker.com/magazine/2010/08/30/face-blind.

10. Sharrona Pearl, "Watching While (Face) Blind: Clone Layering and Prosopagnosia," in *Orphan Black: Performance, Gender, Biopolitics*, ed. Andrea Goulet and Robert A. Rushing (London: Intellect Ltd., 2019), 78–91.

11. Sellers, *You Don't Look Like Anyone I Know*.

12. Scott Gabriel Knowles, Sharrona Pearl, and Rashawn Ray, "COVID Mask Wearing: Identity and Materiality," *East Asian Science, Technology and Society: An International Journal* 16, no. 1 (January 2, 2022): 117–23, https://doi.org/10.1080/18752160.2021.2015134; Sharrona Pearl, "Masked and Anonymous," *Real Life*, July 18, 2016, https://reallifemag.com/masked-and-anonymous/; Sharrona Pearl, "Perspective | The Reason We're So Uncomfortable Wearing Masks," *Washington Post*, May 7, 2020, https://www.washingtonpost.com/outlook/2020/05/07/reason-were-so-uncomfortable-wearing-masks-may-surprise-you/.

Chapter 1. Thinking in Cases

1. Physiognomy, the study of facial features and their relationship to character, was in some ways a form of people differentiation. Though it was not exactly face recognition, it

holds a place in the development of theories around how people made sense of one another, as we see in chapter 5. For more on physiognomy see Sharrona Pearl, *About Faces: Physiognomy in Nineteenth-Century Britain* (Cambridge, MA: Harvard University Press, 2010).

2. I draw here from historian of medicine Monica Green, who has argued forcefully for using the tools available, including DNA evidence, to make sense of past medical and physical conditions while always being attentive to cultural and contextual factors. See also Jonathan Sadowsky, who has brilliantly navigated the question of retrospective diagnoses in his work on depression, demonstrating the advantages of access to greater understanding that a careful approach to past conditions might yield. Monica H. Green, "The Four Black Deaths," *American Historical Review* 125, no. 5 (December 29, 2020): 1601–31, https://doi.org/10.1093/ahr/rhaa511; Jonathan Sadowsky, *The Empire of Depression: A New History*, 1st ed. (Cambridge, UK: Polity, 2020).

3. For the classic text on this topic, see John Forrester, "If p, Then What? Thinking in Cases," *History of the Human Sciences* 9, no. 3 (August 1, 1996): 1–25, https://doi.org/10.1177/095269519600900301. Forrester drew in meaningful ways on Thomas S. Kuhn, *The Structure of Scientific Revolutions*, 3rd ed. (Chicago: University of Chicago Press, 1996). For a recent application of case reasoning as meaningful data, see Stephen Casper, "The Anecdotal Patient: Brain Injury and the Magnitude of Harm," *Notes and Records* 76, no. 4 (February 16, 2022): 663–82, https://royalsocietypublishing.org/doi/10.1098/rsnr.2021.0021; and Sharrona Pearl, "Introduction: Theorizing and Applying the Meaningfully Anecdotal Patient in Neurodiversity Research," *Notes and Records* 76, no. 4 (December 20, 2022): 657–61, https://royalsocietypublishing.org/doi/full/10.1098/rsnr.2021.0083.There are resonances here of Lorraine Daston and Peter Galison's magisterial work on the invention of objectivity and the different ways of deciding what counts as a model from which to reason and know. Lorraine Daston and Peter Galison, *Objectivity* (New York: Zone Books, 2010).

4. Steven G. Epstein, "The Politics of Health Mobilization in the United States: The Promise and Pitfalls of 'Disease Constituencies,'" *Social Science and Medicine* 165 (September 1, 2016): 246–54, https://doi.org/10.1016/j.socscimed.2016.01.048.

5. The dismissal of individual cases is particularly acute in the context of racial bias in US health care, in which patients of color and especially Black patients are dismissed without being believed. Women are also disbelieved at a much higher rate than men. Kelly M. Hoffman et al., "Racial Bias in Pain Assessment and Treatment Recommendations, and False Beliefs about Biological Differences between Blacks and Whites," *Proceedings of the National Academy of Sciences* 113, no. 16 (April 19, 2016): 4296–4301, https://doi.org/10.1073/pnas.1516047113; Verónica Zaragovia, "Trying to Avoid Racist Health Care, Black Women Seek Out Black Obstetricians," NPR, May 28, 2021, https://www.npr.org/sections/health-shots/2021/05/28/996603360/trying-to-avoid-racist-health-care-black-women-seek-out-black-obstetricians; "Bias in Medicine," *Last Week Tonight with John Oliver*, HBO, 2019, https://www.youtube.com/watch?v=TATSAHJKRd8.

6. Ian Hacking, "Styles of Scientific Thinking," in *Post-Analytic Philosophy*, ed. John Rajchman and Cornel West (New York: Columbia University Press, 1985), 145–65; Forrester, "If p, Then What?," 14.

7. Chris Millard and Felicity Callard, "Thinking in, with, across, and beyond Cases with John Forrester," *History of the Human Sciences* 33, no. 3–4 (October 1, 2020): 3–14, https://doi.org/10.1177/0952695120965403.

8. Michigan State Medical Society, *Transactions of the Michigan State Medical Society* 23 (1899), https://www.google.com/books/edition/Transactions_of_the_Michigan_State _Medic/o97hAAAAMAAJ?hl=en&gbpv=1&dq=%22utterly+unable+to+distinguish+some, +or+in+rare+instances,+all+colors.%22&pg=PA379&printsec=frontcover.

9. A. L. Wigan, "The Duality of the Mind, Proved by the Structure, Functions, and Diseases of the Brain," *The Lancet* 43, no. 1074 (March 30, 1844): 39–41, https://www.sciencedirect .com/science/article/abs/pii/S0140673602350396.

10. For more on modern interest in the duality of mind, see Anne Harrington, review of "Arthur Wigan and the Duality of the Mind, Psychological Medicine Monograph Supplement 11" by Basil Clarke, *Psychological Medicine* 19, no. 1 (February 1989): 245–47, https://doi .org/10.1017/S0033291700001211; Robert Louis Stevenson, *The Strange Case of Dr. Jekyll and Mr. Hyde, With Other Fables* (Liberia: Palala Press, 2015).

11. Wigan, "The Duality of the Mind," 171.

12. Wigan, "The Duality of the Mind"; Sergio Della Sala and Andrew W. Young, "Quaglino's 1867 Case of Prosopagnosia," *Cortex* 39, no. 3 (2003): 533–40, https://pubmed.ncbi .nlm.nih.gov/12870826/; Michigan State Medical Society, *Transactions of the Michigan State Medical Society*; Hadyn D. Ellis and Melanie Florence, "Bodamer's (1947) Paper on Prosopagnosia," *Cognitive Neuropsychology* 7, no. 2 (March 1, 1990): 81–105, https://doi.org/10 .1080/02643299008253437; H. Hécaen and R. Angelergues, "Agnosia for Faces (Prosopagnosia)," *Archives of Neurology (Chicago)* 7, no. 2 (1962): 92–100, https://doi.org/10.1001 /archneur.1962.04210020014002; A. L. Benton and M. W. Van Allen, "Impairment in Facial Recognition in Patients with Cerebral Disease," *Transactions of the American Neurological Association* 93 (1968): 38–42; H. R. McConachie, "Developmental Prosopagnosia. A Single Case Report," *Cortex* 12, no. 1 (March 1976): 76–82, https://doi.org/10.1016/s0010 -9452(76)80033-0; M. R. Polster and S. Z. Rapcsak, "Representations in Learning New Faces: Evidence from Prosopagnosia," *Journal of the International Neuropsychological Society: JINS* 2, no. 3 (May 1996): 240–48, https://doi.org/10.1017/s1355617700001181.

13. Sala and Young, "Quaglino's 1867 Case of Prosopagnosia."

14. Sala and Young, 533.

15. Sala and Young, 533.

16. Sala and Young, 534.

17. Paul Eling and Stanley Finger, "Franz Joseph Gall on the Cerebellum as the Organ for the Reproductive Drive," *Frontiers in Neuroanatomy* 13 (2019): 40, https://doi.org/10.3389 /fnana.2019.00040; Sala and Young, "Quaglino's 1867 Case of Prosopagnosia," 535.

18. Sala and Young, "Quaglino's 1867 Case of Prosopagnosia," 534.

19. John Hughlings Jackson, "Case of Disease of the Brain—Left Hemiplegia—Mental Affection," *Medical Times and Gazette*, May 4, 1872, 513–14.

20. Sala and Young, "Quaglino's 1867 Case of Prosopagnosia," 536.

21. Sala and Young, 536; Hughlings Jackson, "Case of Disease of the Brain."

22. Sala and Young, "Quaglino's 1867 Case of Prosopagnosia," 537.

23. Georges Didi-Huberman, *Invention of Hysteria: Charcot and the Photographic Iconography of the Salpêtrière*, trans. Alisa Hartz (Cambridge, MA: MIT Press, 2004).

24. Sala and Young, "Quaglino's 1867 Case of Prosopagnosia," 537.

25. J. M. Charcot and D. Bernard, "Un cas de suppression brusque et isolée de la vision mentale des signes et des objets (formes et couleurs)," *Le Progrès Médical* 11 (1883): 568–71.

26. "Classic Cases in Neuropsychology Book (1) [2nv80q15vdlk]," 30, https://idoc.pub/documents/classic-cases-in-neuropsychology-book-1-2nv80q15vdlk.

27. Charcot and Bernard, "Un cas de suppression."

28. Sala and Young, "Quaglino's 1867 Case of Prosopagnosia," 537.

29. Sacks, "Face-Blind," *New Yorker*, August 23, 2010, https://www.newyorker.com/magazine/2010/08/30/face-blind.

30. Sala and Young, "Quaglino's 1867 Case of Prosopagnosia," 537–38.

31. Sala and Young, 537.

32. Michigan State Medical Society, *Transactions of the Michigan State Medical Society*.

33. There are numerous examples of using the same label for a condition across time even as the understanding of that condition changes. We can consider diseases like syphilis and Crutzfeldt-Jakob that shift from being mental conditions to pathological infections; Down's Syndrome that shifts from being an atavistic condition to a chromosomal one; measles and leprosy as general terms for skin conditions that become specific diseases; hysteria as wandering womb that later becomes subsumed into a host of other conditions; and even the humble cold, as outlined in Olivia Weisser, *Ill Composed: Sickness, Gender, and Belief in Early Modern England*, 1st ed. (New Haven, CT: Yale University Press, 2015).

34. Michigan State Medical Society, *Transactions of the Michigan State Medical Society*, 379.

35. Michigan State Medical Society, 380.

36. Mark Solms, Karen Kaplan-Solms, and Jason W. Brown, "Wilbrand's Case of 'Mind-Blindness,'" in *Classic Cases in Neuropsychology* (Hove, UK: Psychology Press, 1996), 83.

37. Otto Potzl, *Die Aphasielehre vom Standpunkt der klinischen Psychiatrie:* Vol. 1. *Die optisch-agnostischen Störungen (Die verschiedenen Formen der Seelenblindheit)* (Leipzig: Deuticke, 1928).

38. Annemarie Jutel, *Diagnosis: Truths and Tales* (Toronto: University of Toronto Press, 2019); Epstein, "The Politics of Health Mobilization in the United States"; Rayna Rapp and Faye D. Ginsburg, "Enabling Disability: Rewriting Kinship, Reimagining Citizenship," *Public Culture* 13, no. 3 (September 1, 2001): 533–56; Ian Hacking, "The Looping Effects of Human Kinds," in *Causal Cognition: A Multidisciplinary Debate*, 351–94, Symposia of the Fyssen Foundation (New York: Clarendon Press/Oxford University Press, 1995).

39. Ellis and Florence, "Bodamer's (1947) Paper on Prosopagnosia." Note that I draw here from a translation rather than the original.

40. Ellis and Florence, 82.

41. Ellis and Florence, 81.

42. Ellis and Florence, 83.

43. Ellis and Florence, 83.

44. Ellis and Florence, 83.

45. Ellis and Florence, 83.

46. The backdrop of World War II may well have slowed down the global reception of Bodamer's work.

47. For more on the history and science of color, see Michael Rossi, *The Republic of Color: Science, Perception, and the Making of Modern America*, 1st ed. (Chicago: University of Chicago Press, 2019).

48. Ellis and Florence, "Bodamer's (1947) Paper on Prosopagnosia," 84.

49. Ellis and Florence, 84–85.

50. Ellis and Florence, 85.

51. Ellis and Florence, 86.

52. Ellis and Florence, 86.

53. Ellis and Florence, 86.

54. Ellis and Florence, 87.

55. Ellis and Florence, 87.

56. Ellis and Florence, 87.

57. Ellis and Florence, 88.

58. Ellis and Florence, 88.

59. Ellis and Florence, 90.

60. Ellis and Florence, 90–91.

61. Ellis and Florence, 91.

62. Ellis and Florence, 92.

63. Ellis and Florence, 92.

64. Ellis and Florence, 92.

65. Ellis and Florence, 93.

66. Ellis and Florence, 93–94.

67. Ellis and Florence, 94.

68. Ellis and Florence, 95.

69. Ellis and Florence, 94.

70. Ellis and Florence, 96.

71. See, for example, Joseph Dumit, *Picturing Personhood* (Princeton, NJ: Princeton University Press, 2004); Katja Guenther, *Localization and Its Discontents: A Genealogy of Psychoanalysis and the Neuro Disciplines* (Chicago: University of Chicago Press, 2015); Anne Harrington, *Mind Fixers: Psychiatry's Troubled Search for the Biology of Mental Illness*, 1st ed. (New York: W. W. Norton, 2019).

72. Benton and Van Allen, "Impairment in Facial Recognition in Patients with Cerebral Disease"; McConachie, "Developmental Prosopagnosia. A Single Case Report."

73. Polster and Rapcsak, "Representations in Learning New Faces."

74. Benton and Van Allen, "Impairment in Facial Recognition in Patients with Cerebral Disease," 344.

75. Benton and Van Allen.

76. McConachie, "Developmental Prosopagnosia. A Single Case Report"; Elizabeth K. Warrington and Merle James, "An Experimental Investigation of Facial Recognition in Patients with Unilateral Cerebral Lesions," *Cortex* 3, no. 3 (1967): 317–26, https://doi.org/10.1016/S0010-9452(67)80020-0; E. De Renzi, P. Faglioni, and H. Spinnler, "The Performance of Patients with Unilateral Brain Damage on Face Recognition Tasks," *Cortex* 4, no. 1 (March 1, 1968): 17–34, https://doi.org/10.1016/S0010-9452(68)80010-3.

Chapter 2. The Blindness of Great Men

1. For more on diagnosis see Heather Sellers, *You Don't Look Like Anyone I Know: A True Story of Family, Face Blindness, and Forgiveness* (New York: Riverhead, 2011).

2. See, for example, Annemarie Jutel, *Diagnosis: Truths and Tales* (Toronto: University of Toronto Press, 2019); Faith Wallis, "Signs and Senses: Diagnosis and Prognosis in Early Medieval Pulse and Urine Texts," *Social History of Medicine* 13, no. 2 (December 1, 2000): 265–78,

https://doi.org/10.1093/shm/13.2.265; R. Wittern-Sterzel, "Diagnosis: The Doctor and the Urine Glass," *The Lancet* 354 Suppl (December 1999): SIV13, https://doi.org/10.1016 /s0140-6736(99)90356-2; Steven G. Epstein, "The Politics of Health Mobilization in the United States: The Promise and Pitfalls of 'Disease Constituencies,'" *Social Science and Medicine* 165 (September 1, 2016): 246–54, https://doi.org/10.1016/j.socscimed.2016.01.048; Rayna Rapp and Faye D. Ginsburg, "Enabling Disability: Rewriting Kinship, Reimagining Citizenship," *Public Culture* 13, no. 3 (September 1, 2001): 533–56.

　　3. Hadyn D. Ellis and Melanie Florence, "Bodamer's (1947) Paper on Prosopagnosia," *Cognitive Neuropsychology* 7, no. 2 (March 1, 1990): 81–105, https://doi.org/10.1080 /02643299008253437; H. Hécaen and R. Angelergues, "Agnosia for Faces (Prosopagnosia)," *Archives of Neurology (Chicago)* 7, no. 2 (1962): 92–100, https://doi.org/10.1001/archneur.1962 .04210020014002; A. L. Benton and M. W. Van Allen, "Impairment in Facial Recognition in Patients with Cerebral Disease," *Transactions of the American Neurological Association* 93 (1968): 38–42, https://pubmed.ncbi.nlm.nih.gov/5711050/; Elizabeth K. Warrington and Merle James, "An Experimental Investigation of Facial Recognition in Patients with Unilateral Cerebral Lesions," *Cortex* 3, no. 3 (1967): 317–26, https://doi.org/10.1016/S0010 -9452(67)80020-0; E. De Renzi, P. Faglioni, and H. Spinnler, "The Performance of Patients with Unilateral Brain Damage on Face Recognition Tasks," *Cortex* 4, no. 1 (March 1, 1968): 17–34, https://doi.org/10.1016/S0010-9452(68)80010-3.

　　4. H. R. McConachie, "Developmental Prosopagnosia. A Single Case Report," *Cortex* 12, no. 1 (March 1976): 76–82, https://doi.org/10.1016/s0010-9452(76)80033-0.

　　5. Thomas Kress and Irene Daum, "Developmental Prosopagnosia: A Review," *Behavioural Neurology* 14, no. 3–4 (2003): 109–21, https://doi.org/10.1155/2003/520476.

　　6. This footnote is a reward for those who read them. The first time I heard the term "teleological" was in a supervision with Simon Schaffer. He read my dissertation chapter on Hugh Welch Diamond, looked over the paper at me, and told me that as it stood, it was "worryingly teleological." I vowed to never write anything worryingly teleological again, so rest assured I am about to complicate this story.

　　7. Oliver Sacks, "Face-Blind," *The New Yorker*, August 23, 2010, https://www.newyorker .com/magazine/2010/08/30/face-blind.

　　8. Oliver Sacks, *A Leg to Stand On* (New York: Touchstone, 1998); Oliver Sacks, *Migraine*, Rev. ed. (New York: Vintage, 1999).

　　9. Oliver Sacks, *On the Move: A Life* (New York: Vintage, 2016); "The Tragic Story of Oliver Sacks's Celibacy," *Washington Post*, https://www.washingtonpost.com/news/morning -mix/wp/2015/08/31/the-tragic-story-of-oliver-sackss-celibacy-and-homosexuality/.

　　10. Robin Williams and Robert De Niro, *Awakenings* (Culver City, CA: Columbia Pictures, 1990).

　　11. Oliver Sacks, *Awakenings* (New York: Vintage, 1999).

　　12. Sacks, *A Leg to Stand On*.

　　13. Oliver Sacks, *The Man Who Mistook His Wife for a Hat: And Other Clinical Tales* (New York: Touchstone, 1998).

　　14. Anatole Broyard, "Good Books about Being Sick," *New York Times*, April 1, 1990, https://www.nytimes.com/1990/04/01/books/good-books-abut-being-sick.html.

　　15. See, for example, Lawrence Weschler, *And How Are You, Dr. Sacks?* 1st ed. (New York: Farrar, Straus and Giroux, 2019).

16. Sara Hendren, "Avoiding the High-Brow Freak Show," July 20, 2017, https://sarahendren.com/2017/07/20/avoiding-the-high-brow-freak-show/; Tom Shakespeare, "Book Reviews," *Disability & Society* 11, no. 1 (March 1, 1996): 136–41, https://doi.org/10.1080/09687599650023416; Jenny Diski, "Life, Death and the Whole Damn Thing," *London Review of Books*, October 17, 1996, https://www.lrb.co.uk/the-paper/v18/n20/jenny-diski/life-death-and-the-whole-damn-thing.

17. Hendren, "Avoiding the High-Brow Freak Show"; Diski, "Life, Death and the Whole Damn Thing."

18. Shakespeare, "Book Reviews," 136.

19. Leonard Cassuto, "Oliver Sacks: The P. T. Barnum of the Postmodern World?," *American Quarterly* 52, no. 2 (June 2000): 326–33, https://www.jstor.org/stable/30041844.

20. Shakespeare, "Book Reviews"; Sarah Zhang, "Temple Grandin on How Oliver Sacks Changed Her Life," *Wired*, September 1, 2015, https://www.wired.com/2015/09/temple-grandin-oliver-sacks-changed-life/.

21. "Oliver Sacks," *Charlie Rose*, February 23, 1995, https://charlierose.com/videos/5565.

22. Sacks, *On the Move*.

23. Oliver Sacks, "Opinion | My Own Life," *New York Times*, February 19, 2015, https://www.nytimes.com/2015/02/19/opinion/oliver-sacks-on-learning-he-has-terminal-cancer.html.

24. Rosemarie Garland-Thomson, "Opinion | Becoming Disabled," *New York Times*, August 19, 2016, https://www.nytimes.com/2016/08/21/opinion/sunday/becoming-disabled.html.

25. I thank Michael Gordin for this eloquent phrasing.

26. Sacks, *Migraine*.

27. Sacks, *A Leg to Stand On*.

28. Sacks, "Face-Blind."

29. Sacks, *The Man Who Mistook His Wife for a Hat*, 8–22.

30. Sacks, *A Leg to Stand On*.

31. Sacks, "Face-Blind."

32. Sacks, "Face-Blind."

33. Sacks, "Face-Blind."

34. Sacks, "Face-Blind."

35. See, for example, Kelly Oakes, "Are You Face-Blind?," BuzzFeed, May 16, 2017, https://www.buzzfeed.com/kellyoakes/how-face-blind-are-you. Neuroscience explanations are extremely powerful, as Deena Weisberg has shown. Deena Skolnick Weisberg et al., "The Seductive Allure of Neuroscience Explanations," *Journal of Cognitive Neuroscience* 20, no. 3 (March 2008): 470–77, https://doi.org/10.1162/jocn.2008.20040.

36. Steve Bradt, "'Face-Blindness' Disorder May Not Be So Rare," *Harvard Gazette* (blog), June 1, 2006, https://news.harvard.edu/gazette/story/2006/06/face-blindness-disorder-may-not-be-so-rare/.

37. Carey Goldberg, "When Faces Have No Name," *Boston Globe*, June 14, 2006, http://archive.boston.com/news/local/articles/2006/06/14/when_faces_have_no_name/.

38. Much of the historiography has focused on the stigmatizing effect of diagnoses and labeling; here I look at moments when labels offer a valuable framework. For more on this question, see Heather Murray, "Diagnosing the 'Master Mechanism of the Universe' in

Interwar and War-Era America," *Journal of the History of the Behavioral Sciences* 58, no. 2 (2022): 147–62, https://doi.org/10.1002/jhbs.22148; Rachel Louise Moran, "Spitting on My Sources: Depression, DNA, and the Ambivalent Historian," *Journal of the History of the Behavioral Sciences* 58, no. 4 (July 15, 2022), 449–58, https://doi.org/10.1002/jhbs.22219.

39. Sacks, "Face-Blind."

40. Sacks, "Face-Blind."

41. Sacks, "Face-Blind."

42. Richard Cook and Federica Biotti, "Developmental Prosopagnosia," *Current Biology* 26, no. 8 (April 2016): R312–13, https://doi.org/10.1016/j.cub.2016.01.008.

43. Sacks, "Face-Blind."

44. "Prosopagnosia Research at Bournemouth University—Face Blindness Overview," https://prosopagnosiaresearch.org/face-blindness/overview.

45. Heather Sellers, "Accurately Diagnosing Prosopagnosia," *Psychology Today*, December 1, 2010, http://www.psychologytoday.com/blog/you-don-t-look-anyone-i-know/201012/accurately-diagnosing-prosopagnosia.

46. Almudena Toral, "Faceless," *New York Times*, December 26, 2011, https://www.nytimes.com/video/health/100000000958226/faceless.html.

47. Karen Barrow, "Have We Met? Tracing Face Blindness to Its Roots," *New York Times*, December 26, 2011, https://www.nytimes.com/2011/12/27/health/views/face-and-voice-recognition-may-be-linked-in-the-brain-research-suggests.html.

48. "Face Blindness: When Everyone Is a Stranger," *60 Minutes*, CBS News, February 28, 2016, https://www.cbsnews.com/news/60-minutes-face-blindness-when-everyone-is-a-stranger/.

49. Sacks, "Face-Blind."

50. Jenny Edkins, *Face Politics* (Abingdon, UK: Routledge, 2015), 126.

51. Emmeline May, "So Tell Me Who Are You? |," April 6, 2015, https://rockstardinosaurpirateprincess.wordpress.com/2015/04/06/so-tell-me-who-are-you/; for more on May, see Sharrona Pearl, "Watching While (Face) Blind: Clone Layering and Prosopagnosia," in *Orphan Black: Performance, Gender, Biopolitics*, ed. Andrea Goulet and Robert A. Rushing (London: Intellect Ltd., 2019), 78–91.

52. Daniel S Hamermesh, *Beauty Pays: Why Attractive People Are More Successful* (Princeton, NJ: Princeton University Press, 2011); Sharrona Pearl, *Face/On: Face Transplants and the Ethics of the Other* (Chicago: University of Chicago Press, 2017).

53. Goldberg, "When Faces Have No Name."

54. "Face Blindness, Part 1," *60 Minutes*, CBS News, April 21, 2015, https://www.youtube.com/watch?v=dxqsBk7Wn-Y&feature=youtu.be.

55. Angela Epstein, "You Hug Strangers and Can't Recognise Your Husband: The Bizarre Nightmare of 'Faceblindness,'" *Daily Mail*, January 8, 2008, https://www.dailymail.co.uk/health/article-506672/You-hug-strangers-recognise-husband-The-bizarre-nightmare-faceblindness.html.

56. "Face Blindness, Part 1."

57. Epstein, "You Hug Strangers and Can't Recognise Your Husband."

58. Epstein.

59. Sellers, *You Don't Look Like Anyone I Know*.

60. Toral, "Faceless."

61. For details on the allegations against Close see Robin Pogrebin, "Chuck Close Apologizes after Accusations of Sexual Harassment," *New York Times*, December 20, 2017, https://www.nytimes.com/2017/12/20/arts/design/chuck-close-sexual-harassment.html. I take such allegations seriously and am concerned about elevating someone who has admitted to this behavior. I draw here from Rachel Hope Cleves's work on writing about sexual predators to situate Close as a person whose behaviors we resist, while still acknowledging the role he has played and the work he has done. This work is intertwined inextricably with the accusations against him, behaviors to which he has admitted and which he situates in his own artistic practice. I say this not as a caveat for Close's violence; it is inseparable from his art and his platform. It is that very platform that played a role in the prosopagnosia story. Something about the power of Chuck Close looks a lot like the power of a lot of men to determine the terms of their stories. But here, I am the one telling the story. And I say the sexual predation must be included. Cleves has given us some guidance on how to write about sexual predators while being attentive to historically shifting attitudes around acceptable and abhorrent behavior. Rachel Hope Cleves, *Unspeakable: A Life beyond Sexual Morality* (Chicago: University of Chicago Press, 2020).

62. Tom Valeo, "Noted Photographer Chuck Close Transcends His Face Blindness," *Tampa Bay Times*, January 22, 2013, https://www.tampabay.com/things-to-do/visualarts/noted-photographer-chuck-close-transcends-his-face-blindness/1271738/.

63. Society for Neuroscience, *Dialogues between Neuroscience and Society: Neuroscience 2012*, YouTube, 2012, https://www.youtube.com/watch?v=qWadiloW5GU.

64. Todd Farley, "Disabilities Are at the Heart of Chuck Close's Art," *Brain&Life*, August/September 2011, https://www.brainandlife.org/articles/dyslexia-paralysis-face-blindness-nothing-comes-between-legendary-artist-chuck/.

65. World Science Festival, *Strangers in the Mirror*, YouTube, 2015, https://www.youtube.com/watch?v=7OKC9CJoLdk.

66. Sacks, "Face-Blind."

67. Evan Kindley, *Questionnaire* (New York: Bloomsbury Academic, 2016).

68. "Prosopagnosia Research—Home," https://prosopagnosiaresearch.org/; "Prosopagnosia Research at Bournemouth University—Face Blindness Overview," https://prosopagnosiaresearch.org/face-blindness/overview; "Face Recognition," https://faculty.washington.edu/chudler/java/faces.html.

69. Ebony Murray and Sarah Bate, "Diagnosing Developmental Prosopagnosia: Repeat Assessment Using the Cambridge Face Memory Test," *Royal Society Open Science* 7, no. 9 (September 23, 2020): 200884, https://doi.org/10.1098/rsos.200884; Joseph M. Arizpe et al., "Self-Reported Face Recognition Is Highly Valid, but Alone Is Not Highly Discriminative of Prosopagnosia-Level Performance on Objective Assessments," *Behavior Research Methods* 51, no. 3 (June 1, 2019): 1102–16, https://doi.org/10.3758/s13428-018-01195-w.

70. Goldberg, "When Faces Have No Name."

71. Brad Duchaine and Ken Nakayama, "The Cambridge Face Memory Test: Results for Neurologically Intact Individuals and an Investigation of Its Validity Using Inverted Face Stimuli and Prosopagnosic Participants," *Neuropsychologia* 44, no. 4 (January 1, 2006): 576–85, https://doi.org/10.1016/j.neuropsychologia.2005.07.001.

72. Mirta Stantic et al., "The Oxford Face Matching Test: A Non-Biased Test of the Full Range of Individual Differences in Face Perception," *Behavior Research Methods*, June 15, 2021, https://doi.org/10.3758/s13428-021-01609-2.

73. Bradley Duchaine, email to the author, July 11, 2021.

74. James W. Tanaka, Markus Kiefer, and Cindy M. Bukach, "A Holistic Account of the Own-Race Effect in Face Recognition: Evidence from a Cross-Cultural Study," *Cognition* 93, no. 1 (August 1, 2004): B1–9, https://doi.org/10.1016/j.cognition.2003.09.011; David J. Robertson et al., "Super-Recognisers Show an Advantage for Other Race Face Identification," *Applied Cognitive Psychology* 34, no. 1 (2020): 205–16, https://doi.org/10.1002/acp.3608; Hoo Keat Wong et al., "The Other-Race Effect and Holistic Processing across Racial Groups," *Scientific Reports* 11, no. 1 (April 19, 2021): 8507, https://doi.org/10.1038/s41598 -021-87933-1.

75. The role of holistic processing remains the subject of some debate among face recognition researchers. Ahmed M. Megreya, "Feature-by-Feature Comparison and Holistic Processing in Unfamiliar Face Matching," *PeerJ* 6 (2018): e4437, https://doi.org/10.7717/peerj .4437; Christian Østergaard Knudsen, Katrine Winther Rasmussen, and Christian Gerlach, "Gender Differences in Face Recognition: The Role of Holistic Processing," *Visual Cognition* 29, no. 6 (July 3, 2021): 379–85, https://doi.org/10.1080/13506285.2021.1930312.

76. Stantic et al., "The Oxford Face Matching Test."

77. Arizpe et al., "Self-Reported Face Recognition Is Highly Valid."

78. See, for example, Joshua Davis, "Face Blind," *Wired*, November 1, 2006, https://www .wired.com/2006/11/blind/; Ed Yong, "Faulty Connections Responsible for Inherited Face-Blindness," *National Geographic*, November 24, 2008, https://www.nationalgeographic.com /science/article/faulty-connections-responsible-for-inherited-face-blindness; Barrow, "Have We Met?"; "Heather Sellers' Battle With Face Blindness," ABC News, October 18, 2010, https://abcnews.go.com/GMA/Books/video/heather-sellers-battle-face-blindness -11906989; "Face Blindness, Part 1."

79. See, for example, Sellers, *You Don't Look Like Anyone I Know*. There has also been a rise in discussions around face-blind celebrities including Brad Pitt and Swedish Crown Princess Victoria. For a more detailed list from 2013, see "Photos: The Faces of Those Who Don't Recognize Faces," CNN, May 23, 2013, https://www.cnn.com/2013/05/23/health /gallery/notable-prosopagnosia/index.html.

80. There is a small irony in Brad Pitt, who does not recognize faces, being routinely included in tests for face blindness.

81. But, okay: by tracking search terms on Google, there is a major spike of interest in 2006 for "face blindness," "prosopagnosia," and related terms. There is a similar effect for "face blindness test" and "face blindness quiz" as both general search terms and specifically in media/newspapers. This tracks alongside the publication of the initial *Boston Globe* article and the semi-public release of the CFMT. There is another peak in 2012, likely as a result of Oliver Sacks's writing and the related subsequent media attention to prosopagnosia. A Google ngram search shows a steady rise in the term "face blindness" from 2009 to 2019, and an Internet Archive text search shows a major spike in "face blindness" for 2007 to 2015. There is a particular peak in 2010 for tv news captions, and a similar one for radio transcripts in 2017.

82. Gabrielle Sanchez, "A Complete Timeline of Every Joss Whedon Controversy," Vulture, May 12, 2021, https://www.vulture.com/2021/05/complete-timeline-joss-whedon

-allegations.html; Lila Shapiro, "The Undoing of Joss Whedon," Vulture, January 17, 2022, https://www.vulture.com/article/joss-whedon-allegations.html.

83. There is a rich literature on Buffy studies; see, for example, Irene Karras, "The Third Wave's Final Girl: Buffy the Vampire Slayer," *Thirdspace: A Journal of Feminist Theory and Culture*, December 17, 2007, https://journals.lib.sfu.ca/index.php/thirdspace/article/view/karras; Francis H. Early, "Staking Her Claim: Buffy the Vampire Slayer as Transgressive Woman Warrior," *The Journal of Popular Culture* 35, no. 3 (2001): 11–27, https://doi.org/10.1111/j.0022-3840.2001.3503_11.x; Rhonda Wilcox, *Why Buffy Matters: The Art of Buffy the Vampire Slayer* (London: Bloomsbury, 2005).

84. Laura Ruby, *Bone Gap* (New York: Balzer + Bray, 2016); Jennifer Niven, *Holding Up the Universe* (New York: Ember, 2018).

85. Pearl, *Face/On*, 49–87.

86. Ruby, *Bone Gap*.

87. Niven, *Holding Up the Universe*.

88. Ian Hacking, "The Looping Effects of Human Kinds," in *Causal Cognition: A Multidisciplinary Debate*, Symposia of the Fyssen Foundation (New York: Clarendon Press/Oxford University Press, 1995), 351–94.

89. Steven G. Epstein, "Patient Groups and Health Movements," in *The Handbook of Science and Technology Studies* (Cambridge, MA: MIT Press, 2008), 499–539, https://www.scholars.northwestern.edu/en/publications/patient-groups-and-health-movements.

90. Paul Rabinow, "Artificiality and Enlightenment: From Sociobiology to Biosociality," *Politix* 90, no. 2 (2010): 21–46; Joseph Dumit, *Picturing Personhood* (Princeton, NJ: Princeton University Press, 2004), https://press.princeton.edu/titles/7674.html.

Chapter 3. More Men, More Invention

1. Richard Russell, Brad Duchaine, and Ken Nakayama, "Super-Recognizers: People with Extraordinary Face Recognition Ability," *Psychonomic Bulletin & Review* 16, no. 2 (April 2009): 252–57, https://doi.org/10.3758/PBR.16.2.252.

2. Jessica Wolf, "UCLA Internet Studies and Race Scholar Safiya Noble Awarded MacArthur Fellowship," UCLA, September 28, 2021, https://newsroom.ucla.edu/releases/ucla-safiya-noble-macarthur-fellowship.

3. Jak Hutchcraft, "How Police Are Using 'Super Recognizers' Like Me to Track Criminals," Vice, August 24, 2020, https://www.vice.com/en/article/ep487p/how-police-are-using-super-recognizers-like-me-to-track-criminals.

4. For example, Rachel J. Bennetts, Joseph Mole, and Sarah Bate, "Super-Recognition in Development: A Case Study of an Adolescent with Extraordinary Face Recognition Skills," *Cognitive Neuropsychology* 34, no. 6 (August 18, 2017): 357–76, https://doi.org/10.1080/02643294.2017.1402755; "Face Blindness, Part 1," *60 Minutes*, CBS News, April 21, 2015, https://www.youtube.com/watch?v=dxqsBk7Wn-Y&feature=youtu.be

5. Anna K. Bobak, Peter J. B. Hancock, and Sarah Bate, "Super-Recognisers in Action: Evidence from Face-Matching and Face Memory Tasks," *Applied Cognitive Psychology* 30, no. 1 (October 20, 2015): 81–91, https://doi.org/10.1002/acp.3170.

6. Josh P. Davis, "CCTV and the Super-Recognisers," in *Making an Impact on Policing and Crime: Psychological Research, Policy and Practice*, ed. Clifford John T. Stott et al. (London: Routledge, 2020), 34–67.

7. Richard Russell phone interview with the author, October 29, 2018.

8. Russell.

9. Guy Crosby, "Super-Tasters and Non-Tasters: Is It Better to Be Average?," The Nutrition Source, Harvard T. H. Chan School of Public Health, May 31, 2016, https://www.hsph.harvard.edu/nutritionsource/2016/05/31/super-tasters-non-tasters-is-it-better-to-be-average/.

10. Marlene Cimons, "They Never Forget a Face. Research Delves into How 'Super-Recognizers' Can Do This," *Washington Post*, October 30, 2021, https://www.washingtonpost.com/science/super-recognizer-facial-memory/2021/10/29/4cf80caa-2159-11ec-b3d6-8cdebe60d3e2_story.html.

11. Russell phone interview.

12. There was, as Russell noted, certainly a selection bias in the early clinical cases that his team saw, given that they were all self-identified. Russell phone interview

13. Anne McGuire, *War on Autism: On the Cultural Logic of Normative Violence* (Ann Arbor: University of Michigan Press, 2016).

14. Richard Russell's current projects on beauty and makeup are interesting, but his current research is not useful in catching criminals, and none of his current test subjects are going to be known as superheroes. Insights about the subtle differences in, say, face reflectivity as an index to perceptions of beauty are not as interesting to readers of the New Yorker as are stories about super recognizers. Nor is this research as lucrative—but he knew that going in. And he can't quite explain why he chose one path over another except for his sheer fascination with his current research. In a way, it's a quite touching story of passion over pragmatism, intellectual ideals over sensible selection. He just loves what he does, even though the social and academic rewards are so much lower for the beauty work than they would be for super recognition. He told me that if he weren't an atheist, he'd say that God made him do this other kind of work despite the ready-made career opportunities that super recognition offered. It isn't that he had some kind of bad experience or negative encounter around prosopagnosia and super recognition. If anything, it was the kind of heady, invigorating breakthrough story of which movies (and New Yorker articles, and game shows, and intensely gripping academic books) are made.

15. Roeland J. Verhallen et al., "General and Specific Factors in the Processing of Faces," *Vision Research*, "Individual Differences as a Window into the Structure and Function of the Visual System" (special issue) 141 (December 1, 2017): 217–27, https://doi.org/10.1016/j.visres.2016.12.014.

16. James D. Dunn et al., "UNSW Face Test: A Screening Tool for Super-Recognizers," *PLoS ONE* 15, no. 11 (November 16, 2020): e0241747, https://doi.org/10.1371/journal.pone.0241747.

17. Davis, "CCTV and the Super-Recognisers"; Josh Davis interview with the author, October 30, 2018.

18. Davis interview.

19. "Super Recognisers International—The World Leaders in Face Recognition," Super Recognisers International, https://superrecognisersinternational.com/.

20. As Jennifer Jarett exemplifies, not all supers are interested in monetizing their ability or working for law enforcement.

21. Davis interview.

22. For more on this subject, see chapter 6.

23. Meike Ramon, "Super-Recognizers: 70 Cases Identified with a Novel Diagnostic Framework," *Journal of Vision* 21, no. 9 (September 27, 2021): 2354, https://doi.org/10.1167 /jov.21.9.2354; A. Mike Burton, "Why Has Research in Face Recognition Progressed So Slowly? The Importance of Variability," *Quarterly Journal of Experimental Psychology* 66, no. 8 (August 1, 2013): 1467–85, https://doi.org/10.1080/17470218.2013.800125; Meike Ramon, Anna K. Bobak, and David White, "Super-Recognizers: From the Lab to the World and Back Again," *British Journal of Psychology* 110, no. 3 (2019): 461–79, https://doi.org/10.1111/bjop .12368; Russell, Duchaine, and Nakayama, "Super-Recognizers."

24. Hoo Keat Wong, Alejandro J. Estudillo, Ian D. Stephen, and David R. T. Keeble, "The Other-Race Effect and Holistic Processing across Racial Groups," *Scientific Reports* 11, no. 1 (April 19, 2021): 8507, https://doi.org/10.1038/s41598-021-87933-1; David J. Robertson, Jennifer Black, Bethany Chamberlain, Ahmed M. Megreya, and Josh P. Davis, "Super-Recognisers Show an Advantage for Other Race Face Identification," *Applied Cognitive Psychology* 34, no. 1 (2020): 205–16, https://doi.org/10.1002/acp.3608.

25. Davis interview.

26. Davis, "CCTV and the Super-Recognisers." Note that this chapter is itself part of these metrics and was a contribution to the Research Excellence Framework (REF).

27. Cimons, "They Never Forget a Face."

28. Jennifer Jarett, Zoom interview with the author, January 5, 2022.

29. Jarett.

30. Jarett.

31. Linda Hammond, "Super Recognisers," BBC Radio 4, April 6, 2011, https://www.bbc .co.uk/programmes/b00q3fbv; Roni Caryn Rabin, "A Memory for Faces, Extreme Version," *New York Times*, May 25, 2009, https://www.nytimes.com/2009/05/26/health/26face.html; "Face Blindness, Part 1."

32. "Watch Superhuman: Season 1, Episode 5, 'All Parts Extraordinary' Online—FOX," Fox TV promohttps://www.fox.com/watch/492e6e211d567a39dbc4ad07024a10b3.

33. Jarett, Zoom interview.

34. Jarett.

35. Hoo Keat Wong et al., "The Other-Race Effect"; David J. Robertson et al., "Super-Recognisers Show an Advantage"; Alexandra J. Golby, John D. E. Gabrieli, Joan Y. Chiao, and Jennifer L. Eberhardt, "Differential Responses in the Fusiform Region to Same-Race and Other-Race Faces," *Nature Neuroscience* 4, no. 8 (August 2001): 845–50, https://doi.org/10 .1038/90565.

Chapter 4. A Super Useless Super Skill

1. Patrick Radden Keefe, "The Detectives Who Never Forget a Face," *New Yorker*, August 15, 2016, https://www.newyorker.com/magazine/2016/08/22/londons-super-recognizer -police-force.

2. It is quite likely that China has surpassed the United Kingdom in the use and number of CCTV cameras, but that data is not available.

3. Lindybeige, *Super-Recognisers: The Future of Law Enforcement?*, 2017, YouTube, https:// www.youtube.com/watch?v=kQEDzLeAkBs.

4. "The End of the CCTV Era?," *BBC News*, January 15, 2015, http://www.bbc.com/news /magazine-30793614.

5. Lindybeige, *Super-Recognisers*; Rick Noack, "Software Can't Compete with These Cops in Identifying Suspects," *Washington Post*, March 13, 2016, https://www.washingtonpost.com /news/worldviews/wp/2016/03/13/software-cant-compete-with-these-cops-in-identifying -suspects/; Dr. Josh P. Davis, *Super-Recognisers (The One Show, BBC 1 Scotland),* YouTube, 2013, https://www.youtube.com/watch?v=PuPfQ8UZTGQ.

6. Joy Buolamwini, "We Must Fight Face Surveillance to Protect Black Lives," Medium, June 4, 2020, https://onezero.medium.com/we-must-fight-face-surveillance-to-protect -black-lives-5ffcdob4c28a; Joy Buolamwini, "Opinion | When the Robot Doesn't See Dark Skin," *New York Times*, June 22, 2018, https://www.nytimes.com/2018/06/21/opinion/facial -analysis-technology-bias.html; Joy Buolamwini and Timnit Gebru, "Gender Shades: Intersectional Accuracy Disparities in Commercial Gender Classification," *Proceedings of Machine Learning Research* 81 (2018): 77–91, https://proceedings.mlr.press/v81/buolamwini18a.html.

7. For more on the legal issues around face recognition technology, see chapter 6.

8. There are numerous versions of the online super-recognizer test; most are derived from the same basic model, including this one from the University of Greenwich, which has an active super recognizer research lab. "Are You a Super Recogniser?," https:// greenwichuniversity.eu.qualtrics.com/jfe/form/SV_e3xDuCccGAdgbfT; Keefe, "The Detectives Who Never Forget a Face."

9. Anna K. Bobak, Viktoria R. Mileva, and Peter J. B. Hancock, "Facing the Facts: Naive Participants Have Only Moderate Insight into Their Face Recognition and Face Perception Abilities," *Quarterly Journal of Experimental Psychology* 72, no. 4 (April 2019): 872–81, https:// pubmed.ncbi.nlm.nih.gov/29706121/.

10. "Who Is the World's Best Super-Recogniser? This Test Could Help Us Find Them," EurekAlert!, November 16, 2020, https://www.eurekalert.org/news-releases/741625.

11. Heather Sellers, *You Don't Look Like Anyone I Know: A True Story of Family, Face Blindness, and Forgiveness* (New York: Riverhead, 2011); Dan Merica, "John Hickenlooper Didn't Mean to Forget Who You Are: How Face Blindness Has Affected His Political Career," CNN, March 13, 2019, https://www.cnn.com/2019/03/13/politics/john-hickenlooper-face -blindness-prosopagnosia/index.html.

12. Steven G. Epstein, "The Politics of Health Mobilization in the United States: The Promise and Pitfalls of 'Disease Constituencies,'" *Social Science and Medicine* 165 (September 1, 2016): 246–54, https://doi.org/10.1016/j.socscimed.2016.01.048.

13. Steven G. Epstein, "Patient Groups and Health Movements," in *The Handbook of Science and Technology Studies*, 499–539 (Cambridge, MA: MIT Press, 2008), https://www .scholars.northwestern.edu/en/publications/patient-groups-and-health-movements; Karen-Sue Taussig, Rayna Rapp, and Deborah Heath, "Flexible Eugenics: Technologies of the Self in the Age of Genetics," in *Anthropologies of Modernity: Foucault, Governmentality, and Life Politics*, 194–212 (Malden, MA: Blackwell, 2005), https://onlinelibrary.wiley.com/doi/10.1002 /9780470775875.ch8.

14. Sheila Jasanoff, *States of Knowledge: The Co-Production of Science and Social Order* (London: Routledge, 2004).

15. Anne McGuire, *War on Autism: On the Cultural Logic of Normative Violence* (Ann Arbor: University of Michigan Press, 2016); Alison Harnett, "Escaping the Evil Avenger and the

Supercrip: Images of Disability in Popular Television," *Irish Communication Review* 8, no. 1 (November 2, 2016), https://arrow.dit.ie/icr/vol8/iss1/3; Sami Schalk, "Reevaluating the Supercrip," *Journal of Literary & Cultural Disability Studies* 10, no. 1 (2016): 71–86, https://muse.jhu.edu/article/611313/pdf.

16. David T. Mitchell and Sharon L. Snyder, *Narrative Prosthesis: Disability and the Dependencies of Discourse* (Ann Arbor: University of Michigan Press, 2014); Lennard J. Davis, "Constructing Normalcy," in *The Disability Studies Reader*, ed. Lennard J. Davis, 3rd ed. (New York: Routledge, 2010), 9–28; Harnett, "Escaping the Evil Avenger and the Supercrip"; Schalk, "Reevaluating the Supercrip"; Rosemarie Garland Thomson, *Freakery: Cultural Spectacles of the Extraordinary Body* (New York: NYU Press, 1996).

17. Richard Russell, Brad Duchaine, and Ken Nakayama, "Super-Recognizers: People with Extraordinary Face Recognition Ability," *Psychonomic Bulletin & Review* 16, no. 2 (April 2009): 252–57. https://doi.org/10.3758/PBR.16.2.252.

18. Amy Lavole, "'Super-Recognizers' Never Forget a Face," *Harvard Gazette* (blog), May 22, 2009, https://news.harvard.edu/gazette/story/2009/05/super-recognizers-never -forget-a-face/; Keefe, "The Detectives Who Never Forget a Face."

19. Olivia Goldhill, "You're Not a Jerk If You Can't Remember Faces: Facial Blindness Is a Spectrum, Neuroscientists Say," *Quartz* (blog), September 3, 2016, https://qz.com /773522/youre-not-a-jerk-if-you-cant-remember-faces-facial-blindness-is-a-spectrum -neuroscientists-say/.

20. Ian Hacking, "The Looping Effects of Human Kinds," in *Causal Cognition: A Multi-disciplinary Debate*, 351–94, Symposia of the Fyssen Foundation (New York: Clarendon Press/ Oxford University Press, 1995).

21. Epstein, "Patient Groups and Health Movements."

22. "Face Blindness, Part 1," *60 Minutes*, CBS, April 21, 2015, https://www.youtube.com /watch?app=desktop&v=XcgK_7kQUwQ.

23. Marlene Cimons, "They Never Forget a Face. Research Delves into How 'Super-Recognizers' Can Do This," *Washington Post*, October 30, 2021, https://www.washingtonpost .com/science/super-recognizer-facial-memory/2021/10/29/4cf80caa-2159-11ec-b3d6 -8cdebe60d3e2_story.html.

24. "Face Blindness, Part 1"; Roni Caryn Rabin, "A Memory for Faces, Extreme Version," *New York Times*, May 25, 2009, https://www.nytimes.com/2009/05/26/health/26face .html.

25. Rick Noack, "Software Can't Compete with These Cops in Identifying Suspects," *Washington Post*, March 13, 2016, https://www.washingtonpost.com/news/worldviews/wp /2016/03/13/software-cant-compete-with-these-cops-in-identifying-suspects/.

26. Rachel J. Bennetts, Joseph Mole, and Sarah Bate, "Super-Recognition in Development: A Case Study of an Adolescent with Extraordinary Face Recognition Skills," *Cognitive Neuropsychology* 34, no. 4 (2017): 1–20, https://www.researchgate.net/publication/321237653 _Super-recognition_in_development_A_case_study_of_an_adolescent_with_extraordinary _face_recognition_skills; Anna K. Bobak, Benjamin A. Parris, Nicola J. Gregory, Rachel J. Bennetts, and Sarah Bate, "Eye-Movement Strategies in Developmental Prosopagnosia and 'Super' Face Recognition," *The Quarterly Journal of Experimental Psychology* 70, no. 2 (February 1, 2017): 201–17, https://doi.org/10.1080/17470218.2016.1161059; Daniel B. Elbich and Suzanne Scherf, "Beyond the FFA: Brain-Behavior Correspondences in Face Recognition

Abilities," *NeuroImage* 147 (February 15, 2017): 409–22, https://doi.org/10.1016/j.neuroimage.2016.12.042.

27. Face-blind people also have numerous useful skills; they are particularly good at evaluating facial expressions, tone of voice, and body language, though social anxiety tends to limit how often they apply these abilities. Emmeline May, "So Tell Me Who Are You?," April 6, 2015, https://rockstardinosaurpirateprincess.wordpress.com/2015/04/06/so-tell-me-who-are-you/.

28. Lindybeige, *Super-Recognisers*.

29. Benedikt Emanuel Wirth and Claus-Christian Carbon, "An Easy Game for Frauds? Effects of Professional Experience and Time Pressure on Passport-Matching Performance," *Journal of Experimental Psychology: Applied* 23, no. 2 (2017): 138–57, https://doi.org/10.1037/xap0000114; Anna Katarzyna Bobak, Andrew James Dowsett, and Sarah Bate, "Solving the Border Control Problem: Evidence of Enhanced Face Matching in Individuals with Extraordinary Face Recognition Skills," *PLOS ONE* 11, no. 2 (February 1, 2016): e0148148, https://doi.org/10.1371/journal.pone.0148148.

30. Mike Orcutt, "The Face Recognition Systems That Law Enforcement Agencies Use Are Probably Biased," *MIT Technology Review*, July 6, 2016, https://www.technologyreview.com/s/601786/are-face-recognition-systems-accurate-depends-on-your-race/. For more on recent innovations, see chapter 6.

31. Elena Belanova, Josh P. Davis, and Trevor Thompson, "The Part-Whole Effect in Super-Recognisers and Typical-Range-Ability Controls," *Vision Research* 187 (October 1, 2021): 75–84, https://doi.org/10.1016/j.visres.2021.06.004.

32. Noack, "Software Can't Compete with These Cops in Identifying Suspects."

33. "Are You a Super Recogniser?"

34. Reddit is an anonymous and moderated site, so I used the first-person plural for gender designations. Not all of the AMA material here is strictly relevant, so I have grouped the comments based on theme and applicability, and I left the grammatical constructions and typos as I found them to make it easier to read through without interruption.

35. There have been numerous thoughtful discussions and debates around the ethics of citing internet research. My methodological approach is outlined in Sharrona Pearl, "Watching While (Face) Blind: Clone Layering and Prosopagnosia," in *Orphan Black: Performance, Gender, Biopolitics*, edited by Andrea Goulet and Robert A. Rushing, 78–91 (London: Intellect Ltd., 2019), http://www.press.uchicago.edu/ucp/books/book/distributed/O/bo31275041.html, in which I discuss that Reddit, as a public discussion forum, need not be anonymized and can be cited directly. For more, see also Annette Markham and Elizabeth Buchanan, "Recommendations from the AoIR Ethics Working Committee (Version 2.0)," Association of Internet Researchers (2012), 19.

36. Dr. Ashok Jansari and r/Science, "Science AMA Series: Hi Reddit, I'm Dr Ashok Jansari, a Neuropsychologist at Goldsmiths, University of London. I Research Individuals with Face-Blindness and so Called 'Super Recognisers', Who Have an Almost Superhuman Ability to Recognise Faces. AMA!," *The Winnower*, August 12, 2016, https://doi.org/10.15200/winn.147091.19861.

37. Dr. Ashok Jansari and r/Science, "Science AMA Series."

38. Moira Jones, "Day 2 of Digest Super Week: Meet a Super Recogniser," *The Psychologist*, October 8, 2013, https://www.bps.org.uk/research-digest/day-2-digest-super-week.

39. Dr. Ashok Jansari and r/Science, "Science AMA Series."

40. Dr. Ashok Jansari and r/Science.

41. Dr. Ashok Jansari and r/Science.

42. Dr. Ashok Jansari and r/Science.

43. Lindybeige, *Super-Recognisers*.

44. Dr. Josh P. Davis, *Inside Out BBC1 Super-Recognisers (Dr Josh P Davis)*, YouTube, October 19, 2015, https://www.youtube.com/watch?v=TPGf6kDnYeM.

45. Lindybeige, *Super-Recognisers*.

46. Lindybeige.

47. João Medeiros, "How Police Super-Recognisers Cracked the Russian Novichok Case," *Wired*, August 9, 2018, https://www.wired.co.uk/article/salisbury-novichok-poisoning-russia-suspects.

48. Noack, "Software Can't Compete with These Cops in Identifying Suspects."

49. Lindybeige, *Super-Recognisers*.

50. Sarah Bate et al., "Applied Screening Tests for the Detection of Superior Face Recognition," *Cognitive Research: Principles and Implications* 3, no. 1 (June 27, 2018): 22, https://doi.org/10.1186/s41235-018-0116-5.

51. Katrin Bennhold, "London Police 'Super Recognizer' Walks Beat with a Facebook of the Mind," *New York Times*, October 9, 2015, https://www.nytimes.com/2015/10/10/world/europe/london-police-super-recognizer-walks-beat-with-a-facebook-of-the-mind.html.

52. Dr. Ashok Jansari and r/Science, "Science AMA Series."

53. Alexandra J. Golby et al., "Differential Responses in the Fusiform Region to Same-Race and Other-Race Faces," *Nature Neuroscience* 4, no. 8 (August 2001): 845–50, https://doi.org/10.1038/90565.

54. Dr. Ashok Jansari and r/Science, "Science AMA Series."

55. Dr. Ashok Jansari and r/Science.

56. Hougaard Winterbach, "Heroes and Superheroes: From Myth to the American Comic Book," *South African Journal of Art History* 21, no. 1 (January 1, 2006): 114–34, https://repository.up.ac.za/handle/2263/10798.

57. Harnett, "Escaping the Evil Avenger and the Supercrip."

58. "Face Blindness, Part 1"; Keefe, "The Detectives Who Never Forget a Face."

59. Jennifer Jarett, Zoom interview with the author, January 5, 2022.

60. Great Big Story, *I Never Forget a Face*, YouTube, https://www.youtube.com/watch?v=m2ZBL65fSe8; Davis, *Inside Out BBC1*; WalrusRider, *Super Recognisers Catching Criminals—"Eye Catching" (2016)*, YouTube, 2017, https://www.youtube.com/watch?v=oiXLXjVbDIo.

61. WalrusRider, *Super Recognisers Catching Criminals*.

62. WalrusRider.

63. Pearl, "Watching While (Face) Blind: Clone Layering and Prosopagnosia."

64. Jenny Edkins, *Face Politics* (Abingdon, UK: Routledge, 2015).

65. Bennetts, Mole, and Bate, "Super-Recognition in Development."

66. Richard Russell and Pawan Sinha, "Real-World Face Recognition: The Importance of Surface Reflectance Properties," *Perception* 36, no. 9 (September 1, 2007): 1368–74, https://doi.org/10.1068/p5779.

67. Kirsten A. Dalrymple et al., "'A Room Full of Strangers Every Day': The Psychosocial Impact of Developmental Prosopagnosia on Children and Their Families," *Journal of*

Psychosomatic Research 77, no. 2 (August 2014): 144–50, https://doi.org/10.1016/j.jpsychores .2014.06.001.

68. Meike Ramon et al., "Super-Memorizers Are Not Super-Recognizers," *PLOS ONE* 11, no. 3 (March 23, 2016): e0150972, https://doi.org/10.1371/journal.pone.0150972.

69. "Who Is the World's Best Super-Recogniser?"

70. Cimons, "They Never Forget a Face."

Chapter 5. Face Surveillance at the Border

1. Johana Bhuiyan, "'There's Cameras Everywhere': Testimonies Detail Far-Reaching Surveillance of Uyghurs in China," *The Guardian*, September 30, 2021, https://www .theguardian.com/world/2021/sep/30/uyghur-tribunal-testimony-surveillance-china; Subin Paul and David O. Dowling, "Digital Archiving as Social Protest," *Digital Journalism* 6, no. 9 (October 21, 2018): 1239–54, https://doi.org/10.1080/21670811.2018.1493938; Joce Sterman, Alex Brauer, and Andrea Nejman, "Facial Recognition Technology in School Hallways: States Face a Divisive Debate," WPMI, March 1, 2021, https://mynbc15.com/news/spotlight-on -america/facial-recognition-technology-in-school-hallways-states-face-a-divisive-debate. Bhuiyan recounts the use of technology as a form of protest and mobilization. This is only a small snapshot of the countless uses of surveillance as a repressive technology.

2. The historical path I am about to trace is a well-trodden one. I've walked it a bit be-fore in some of my own work, as have other scholars of the face and head including Jessica Helfand, Simone Browne, and Courtney Thompson. Sharrona Pearl, *About Faces: Physiognomy in Nineteenth-Century Britain* (Cambridge, MA: Harvard University Press, 2010); Court-ney E. Thompson, *An Organ of Murder: Crime, Violence, and Phrenology in Nineteenth-Century America*, 1st ed. (New Brunswick, NJ: Rutgers University Press, 2021); Simone Browne, *Dark Matters: On the Surveillance of Blackness* (Durham, NC: Duke University Press, 2015); Jessica Helfand, *Face: A Visual Odyssey* (Cambridge, MA: MIT Press, 2019). For a recent addition to the narrative see Amanda Levendowski, "Face Surveillance Was Always Flawed," *Public Books* (blog), November 30, 2021, https://www.publicbooks.org/face-surveillance-was -always-flawed/.

3. The exhibit is online and can be seen at "Facial Recognition by Wende Museum—Issuu," December 12, 2015, https://issuu.com/wendemuseum/docs/updated_facial_recognition.

4. See, for example, Pearl, *About Faces*; Mary Gibson, *Born to Crime: Cesare Lombroso and the Origins of Biological Criminology* (Westport, CT: Praeger, 2002); Nicholas Wright Gillham, *A Life of Sir Francis Galton: From African Exploration to the Birth of Eugenics*, 1st ed. (New York: Oxford University Press, 2001); Josh Ellenbogen, *Reasoned and Unreasoned Images: The Photog-raphy of Bertillon, Galton, and Marey*, 1st ed. (University Park: Pennsylvania State Univer-sity Press, 2012).

5. Francis Galton, "Composite Photographs," *Nature*, May 23, 1878, 97–100; for the im-ages, see "Francis Galton and Composite Portraiture," https://galton.org/composite.htm.

6. Pearl, *About Faces*.

7. Nicola Twilley, "Out of Many, One," *New Yorker*, August 22, 2014, https://www .newyorker.com/tech/annals-of-technology/out-of-many-one.

8. Michael A. Grodin, Erin L. Miller, and Johnathan I. Kelly, "The Nazi Physicians as Leaders in Eugenics and 'Euthanasia': Lessons for Today," *American Journal of Public Health*

108, no. 1 (January 2018): 53–57, https://doi.org/10.2105/AJPH.2017.304120; Robert A. Wilson, *The Eugenic Mind Project* (Cambridge, MA: MIT Press, 2017).

9. For more on eugenics and identification practices, see Os Keyes, "Automating Autism: Disability, Discourse, and Artificial Intelligence," *Journal of Sociotechnical Critique* 1, no. 1 (December 4, 2020), https://doi.org/10.25779/89bj-j396; Ellen Samuels, *Fantasies of Identification: Disability, Gender, Race* (New York: New York University Press, 2014).

10. "Unwanted Sterilization and Eugenics Programs in the United States," *Independent Lens* (blog), https://www.pbs.org/independentlens/blog/unwanted-sterilization-and-eugenics-programs-in-the-united-states/; "Immigration Detention and Coerced Sterilization: History Tragically Repeats Itself," American Civil Liberties Union, https://www.aclu.org/news/immigrants-rights/immigration-detention-and-coerced-sterilization-history-tragically-repeats-itself/. There are also numerous global examples of eugenics and forced sterilization. For a recent discussion, see Jacqueline Antonovich, "White Coats, White Hoods: The Medical Politics of the Ku Klux Klan in 1920s America," *Bulletin of the History of Medicine* 95, no. 4 (Winter 2021), https://doi.org/10.1353/bhm.2021.0053.

11. Alexandra Reeve Givens, "How Algorithmic Bias Hurts People with Disabilities," *Slate*, February 6, 2020, https://slate.com/technology/2020/02/algorithmic-bias-people-with-disabilities.html; for a grounded example that combines surveillance, AI, and diagnosis, see Dian Hong et al., "Genetic Syndromes Screening by Facial Recognition Technology: VGG-16 Screening Model Construction and Evaluation," *Orphanet Journal of Rare Diseases* 16, no. 1 (August 3, 2021): 344, https://doi.org/10.1186/s13023-021-01979-y.

12. Shaun Raviv, "The Secret History of Facial Recognition," *Wired*, January 21, 2021, https://www.wired.com/story/secret-history-facial-recognition/; Levendowski, "Face Surveillance Was Always Flawed." See also Nikki Stevens and Os Keyes, "Seeing Infrastructure: Race, Facial Recognition and the Politics of Data," *Cultural Studies* 35, no. 4–5 (2021): 833–53, https://doi.org/10.1080/09502386.2021.1895252.

13. M.üge Çarıkçı and Figen Özen, "A Face Recognition System Based on Eigenfaces Method," *Procedia Technology* 1 (2012): 118–23, https://doi.org/10.1016/j.protcy.2012.02.023; Graham M. Davies and Tim Valentine, "Facial Composites: Forensic Utility and Psychological Research," in *The Handbook of Eyewitness Psychology*, vol. 2, ed. Rod Lindsay, David Ross, Don Read, and Michael Toglia (New York: Psychology Press, 2007), 59–85, https://doi.org/10.4324/9781315805535.ch3.

14. David J. Robertson, "Face Recognition: Security Contexts, Super-Recognizers, & Sophisticated Fraud," *HDIAC* 5, no. 1 (Spring 2018): 7–10. Ahmed M. Megreya, "Feature-by-Feature Comparison and Holistic Processing in Unfamiliar Face Matching," *PeerJ* 6 (2018): e4437, https://doi.org/10.7717/peerj.4437.

15. Robertson, "Face Recognition: Security Contexts, Super-Recognizers, & Sophisticated Fraud"; Megreya, "Feature-by-Feature Comparison and Holistic Processing in Unfamiliar Face Matching."

16. Foreigners could also cross at the Friedrichstraße railway station, but members of the allied forces could not.

17. For a list of films set at Checkpoint Charlie, see "Filming Location Matching 'Checkpoint Charlie, Kreuzberg, Berlin, Germany' (Sorted by Popularity Ascending)," IMDb, http://www.imdb.com/search/title/?locations=Checkpoint+Charlie,+Kreuzberg,+Berlin,+Germa

ny. See also Iain MacGregor, *Checkpoint Charlie: The Cold War, The Berlin Wall, and the Most Dangerous Place on Earth* (New York: Scribner, 2019).

18. Marieke Drost, "Nose Matching at Checkpoint Charlie," *Iron Curtain* (blog), https://www.ironcurtainproject.eu/en/stories/nose-matching-at-checkpoint-charlie/.

19. Drost.

20. Drost.

21. "Peter Bochmann Border Guard Collection," Wende Museum, http://www.wendemuseum.org/collections/peter-bochmann-border-guard-collection.

22. "Historical Witness—Peter Bochmann," Wende Museum, http://www.wendemuseum.org/participate/historical-witness-peter-bochmann.

23. "Historical Witness—Peter Bochmann."

24. Natasha Frost, "The History of Passport Photos, from 'Anything Goes' to Today's Mugshots," Atlas Obscura, September 8, 2017, http://www.atlasobscura.com/articles/passport-photos-history-development-regulation-mugshots; John C. Torpey, *The Invention of the Passport: Surveillance, Citizenship and the State*, 2nd ed. (Cambridge: Cambridge University Press, 2018).

25. Raviv, "The Secret History of Facial Recognition."

26. Naphtali Abudarham, Sarah Bate, Brad Duchaine, and Galit Yovel, "Developmental Prosopagnosics and Super Recognizers Rely on the Same Facial Features Used by Individuals with Normal Face Recognition Abilities for Face Identification," *Neuropsychologia* 160 (September 17, 2021): 107963, https://doi.org/10.1016/j.neuropsychologia.2021.107963; Jessica Tardif et al., "Use of Face Information Varies Systematically from Developmental Prosopagnosics to Super-Recognizers," *Psychological Science* 30, no. 2 (February 2019): 300–8, https://doi.org/10.1177/0956797618811338.

27. Megreya, "Feature-by-Feature Comparison and Holistic Processing in Unfamiliar Face Matching."

28. Drost, "Nose Matching."

29. Drost.

30. For more on super voice recognition, see Virginia Aglieri et al., "The Glasgow Voice Memory Test: Assessing the Ability to Memorize and Recognize Unfamiliar Voices," *Behavior Research Methods* 49, no. 1 (February 1, 2017): 97–110, https://doi.org/10.3758/s13428-015-0689-6; Ryan E. Jenkins et al., "Are Super-Face-Recognisers Also Super-Voice-Recognisers? Evidence from Cross-Modal Identification Tasks," *Applied Cognitive Psychology* 35, no. 3 (2021): 590–605, https://doi.org/10.1002/acp.3813.

31. Numerous scholarly fields have grappled with the ethics of bearing witness to wrongdoing either historically or in real time. For recent discussions of this topic in anthropology, see George J. Kunnath, "Anthropology's Ethical Dilemmas: Reflections from the Maoist Fields of India," *Current Anthropology* 54, no. 6 (December 1, 2013): 740–52, https://doi.org/10.1086/673860; Luis A. Vivanco, "Fieldwork Ethics Forum," *Field Notes*, Oxford University Press, https://global.oup.com/us/companion.websites/9780190642198/ethics/; Liana Chua, "Witnessing the Unseen: Extinction, Spirits, and Anthropological Responsibility," *Cambridge Journal of Anthropology* 39, no. 1 (March 1, 2021): 111–29, https://doi.org/10.3167/cja.2021.390108. For a historian's perspective, see Rachel Hope Cleves, *Unspeakable: A Life beyond Sexual Morality* 1st ed. (Chicago: University of Chicago Press, 2020).

32. That's right: Bochmann just packed up the office and left with many of the materials inside it, and during all the changes, no one stopped to wonder what happened to the documents at the checkpoints. I learned this from an email from the Wende Museum director and cite it with permission. Joes Segel, "With Apologies, Another Question" (email to the author), September 21, 2021.

33. Segel, "With Apologies, Another Question."

34. The question of how to use the results of unethical and harmful research continues to be fraught, even as the application of ill-gotten knowledge is robust and ongoing. Notorious examples include the results of prison experimentation, non-consensual medical experimentation of vulnerable populations including prisoners of war, incarcerated people, children, mentally ill people, and people being detained against their will. This includes the torture of people in concentration camps, the Tuskegee syphilis experiments, and experimentation on colonized peoples, among many others.

Bioethicist Arthur Caplan has made a strong case that any research from Nazi torture experiments must be off-limits for any application at all, regardless of any theoretical (and, he insists, dubious and non-reproducible) life-saving potential. Arthur L. Caplan, "How Should We Regard Information Gathered in Nazi Experiments?," *AMA Journal of Ethics* 23, no. 1 (January 1, 2021): 55–58, https://doi.org/10.1001/amajethics.2021.55.

We can extend Caplan's argument to many other examples of harmful, non-consensual, and deeply depraved human experimentation, including the Tuskegee syphilis trials, the Holmesburg Prison studies, the Depo-Provera birth control studies, and countless other profoundly harmful if less well-known cases. Allen M. Hornblum, *Acres of Skin: Human Experiments at Holmesburg Prison*, 1st ed. (New York: Routledge, 1999).

In addition to the clear ethical breaches that render the material itself profoundly suspect, Caplan noted that the starvation and torture of those on whom the experiments were conducted was so profoundly outside the general norm that any of the findings would themselves be basically meaningless. While some might wish to claim that *at least* some possible meaning can be derived by using these horrific experiences to benefit others, there is a kind of comfort in insisting the data themselves are flawed and unusable. But sometimes, that's simply not the case. Sometimes, the experiments were profoundly unethical but the findings valuable. Do we still adhere to Caplan's doctrine that unethical research is fundamentally tainted and thus unusable? What, then, do we do with the literally uncountable applications of Henrietta Lacks's cells, which were obtained without the consent of her and her family? So much historical medical research was conducted without consent and often with harm; is it even possible to disentangle the original work from current methods and applications? Rebecca Skloot, *The Immortal Life of Henrietta Lacks* (New York: Crown, 2010). Rather than lay aside the work, some insist, the intervention must be to acknowledge the processes by which it came about, to name the names of those who enabled violence and oppression individually and on a broad structural level, to note the coercion or abuses involved, and to commit to change going forward. Can we make a distinction between applications of unethically derived data and explicit torturous human experimentation? While that seems like a potentially useful approach, the difference is in degree rather than kind. Open to interpretation are questions of what constitutes torture, and indeed what constitutes human? Answers to those questions have not been a matter of consensus; see for example the practice of enslavement in the United States. The nature of humanity and its boundaries is of course not, and has never been, only about the incursion of technology. Humanity has always been a historically contingent category; in many ways this book is about the role that relationality plays in making humanity itself legible between actors. The work is ongoing, messy, and deeply incomplete. We cannot always answer these questions, but the very asking of them makes us more attentive to the continuing use of unethically derived information. They highlight the ways in which medical and scientific practice as a whole is deeply intertwined and, in some cases, fundamentally embedded in such histories, legacies, and continuing efforts. While we can make good-faith distinctions between torturing people in an experimental context and, say, attempts to improve border surveillance, both leave us with ethically fraught data.

35. "Urgent Action Needed over Artificial Intelligence Risks to Human Rights," UN News, September 15, 2021, https://news.un.org/en/story/2021/09/1099972.

36. For a brilliant critique of current approaches to AI ethics and concrete suggestions for improvements, see Keyes, "Automating Autism."

Chapter 6. Face Recognition Software and Machine Translation

1. Nicholas F. Benson, "Revisiting Carroll's Survey of Factor-Analytic Studies: Implications for the Clinical Assessment of Intelligence," *Psychological Assessment* 30, no. 8 (20180524): 1028, https://doi.org/10.1037/pas0000556; Leila Zenderland, *Measuring Minds: Henry Herbert Goddard and the Origins of American Intelligence Testing* (Cambridge: Cambridge University Press, 2001).

2. See, for example, Ruha Benjamin, *Race After Technology: Abolitionist Tools for the New Jim Code*, 1st ed. (Medford, MA: Polity, 2019); Simone Browne, *Dark Matters: On the Surveillance of Blackness* (Durham, NC: Duke University Press, 2015); Safiya Noble, *Algorithms of Oppression: How Search Engines Reinforce Racism*, 1st ed. (New York: New York University Press, 2018); C. Riley Snorton, *Black on Both Sides: A Racial History of Trans Identity*, 3rd ed. (Minneapolis: University of Minnesota Press, 2017); Toby Beauchamp, *Going Stealth: Transgender Politics and U.S. Surveillance Practices* (Durham, NC: Duke University Press, 2019); Cathy O'Neil, *Weapons of Math Destruction: How Big Data Increases Inequality and Threatens Democracy* (New York: Crown, 2017); Shoshana Zuboff, *The Age of Surveillance Capitalism: The Fight for a Human Future at the New Frontier of Power*, 1st ed. (New York: PublicAffairs, 2019); Meredith Whittaker et al., "Disability, Bias, and AI," AI Now Institute, New York University, November 2019, 32, https://nyuscholars.nyu.edu/en/publications/disability-bias-and-ai; Shalini Kantayya, dir., *Coded Bias*, https://www.codedbias.com.

3. Os Keyes, "The Bones We Leave Behind," *Real Life*, October 7, 2019, https://reallifemag.com/the-bones-we-leave-behind/; for an early set of discussions on this, see also Philip E. Agre, "Your Face Is Not a Bar Code: Arguments Against Automatic Face Recognition in Public Places," *Whole Earth* 106 (September 10, 2003): 74–77, https://pages.gseis.ucla.edu/faculty/agre/bar-code.html.

4. Many programmers continue to try to use computer neural networks to learn about human workings. See, for example, Alan L. Yuille and Chenxi Liu, "Deep Nets: What Have They Ever Done for Vision?," *International Journal of Computer Vision* 129, no. 3 (March 2021): 781–802, https://doi.org/10.1007/s11263-020-01405-z; Lizhen Liang and Daniel E. Acuna, "Artificial Mental Phenomena: Psychophysics as a Framework to Detect Perception Biases

in AI Models," *FAT* '20: Proceedings of the 2020 Conference on Fairness, Accountability, and Transparency* (January 27, 2020): 403–12, https://doi.org/10.1145/3351095.3375623; Nikki Stevens and Os Keyes, "Seeing Infrastructure: Race, Facial Recognition and the Politics of Data," *Cultural Studies* 35, no. 4–5 (2021): 833–53, https://www.tandfonline.com/doi/abs/10.1080/09502386.2021.1895252.

5. Stevens and Keyes, "Seeing Infrastructure"; "Alex 'Sandy' Pentland," MIT Media Lab, https://www.media.mit.edu/people/sandy/overview/.

6. Michael D. Gordin, "The Forgetting and Rediscovery of Soviet Machine Translation," *Critical Inquiry* 46, no. 4 (June 1, 2020): 835–66, https://doi.org/10.1086/709226.

7. Gordin, 836–37.

8. Gordin, 838.

9. "The Limits of Computer Translations," *The Economist*, January 5, 2017, https://www.economist.com/technology-quarterly/2017/01/05/the-limits-of-computer-translations.

10. Gordin, "The Forgetting and Rediscovery of Soviet Machine Translation," 864.

11. Stephanie Dick, "Of Models and Machines: Implementing Bounded Rationality," *Isis*, November 18, 2015, https://doi.org/10.1086/683527.

12. Stephanie Dick, "Artificial Intelligence," *Harvard Data Science Review* 1, no. 1 (July 1, 2019), https://doi.org/10.1162/99608f92.92fe150c.

13. Dick, "Artificial Intelligence"; Luciano Floridi, *The Fourth Revolution: How the Infosphere Is Reshaping Human Reality* (Oxford: Oxford University Press, 2014).

14. Dick, "Artificial Intelligence"; A. Blair and J. Pollack, "What Makes a Good Co-Evolutionary Learning Environment?," *Computer Science*, 1997, https://www.semanticscholar.org/paper/What-Makes-a-Good-Co-Evolutionary-Learning-Blair-Pollack/c29fb6f72f5e3478d2d52c885a5f8d0409066144; Gerald Tesauro, "Temporal Difference Learning and TD-Gammon," *Communications of the ACM* 38, no. 3 (March 1, 1995): 58–68, https://doi.org/10.1145/203330.203343.

15. Gordin, "The Forgetting and Rediscovery of Soviet Machine Translation"; Dick, "Of Models and Machines"; Dick, "Artificial Intelligence."

16. Columbia Data Science Institute, *Race + Data Science Lecture Series: Ali Alkhatib*, YouTube, March 8, 2022, https://www.youtube.com/watch?v=B1v2KIdL5Rs.

17. Will Knight, "What AlphaGo Can Teach Us about How People Learn," *Wired*, December 23, 2020, https://www.wired.com/story/what-alphago-teach-how-people-learn/.

18. Marlene Cimons, "They Never Forget a Face. Research Delves into How 'Super-Recognizers' Can Do This," *Washington Post*, October 30, 2021, https://www.washingtonpost.com/science/super-recognizer-facial-memory/2021/10/29/4cf80caa-2159-11ec-b3d6-8cdebe60d3e2_story.html.

19. For more on eigenfaces, including a detailed explanation, see Stevens and Keyes, "Seeing Infrastructure."

20. M. A. Turk and A. P. Pentland, "Face Recognition Using Eigenfaces," *Proceedings. 1991 IEEE Computer Society Conference on Computer Vision and Pattern Recognition* (1991): 586–91, https://doi.org/10.1109/CVPR.1991.139758.

21. Jason Brownlee, "A Gentle Introduction to Deep Learning for Face Recognition," *Machine Learning Mastery* (blog), May 30, 2019, https://machinelearningmastery.com/introduction-to-deep-learning-for-face-recognition/.

22. Keyes, "The Bones We Leave Behind"; Whittaker et al., "Disability, Bias, and AI"; Clare Garvie, "Garbage In. Garbage Out. Face Recognition on Flawed Data," Georgetown Law Center on Privacy & Technology, May 16, 2019, https://www.flawedfacedata.com.

23. Shaun Raviv, "The Secret History of Facial Recognition," *Wired*, January 21, 2021, https://www.wired.com/story/secret-history-facial-recognition/.

24. Sidney Perkowitz, "The Bias in the Machine," *Nautilus*, August 19, 2020, https://nautil.us/the-bias-in-the-machine-9209/.

25. Perkowitz.

26. Raviv, "The Secret History of Facial Recognition."

27. For a discussion of error rates in face recognition software, see "NIST Study Evaluates Effects of Race, Age, Sex on Face Recognition Software," National Institute of Standards and Technology, December 19, 2019, https://www.nist.gov/news-events/news/2019/12/nist-study-evaluates-effects-race-age-sex-face-recognition-software.

28. Niraj Chokshi, "Facial Recognition's Many Controversies, From Stadium Surveillance to Racist Software," *New York Times*, May 15, 2019, https://www.nytimes.com/2019/05/15/business/facial-recognition-software-controversy.html. These nineteen people were not arrested, and the unpublicized use of the technology led to protests by the ACLU and others once it was revealed. "Use of Facial Recognition at Super Bowl and in Tampa," American Civil Liberties Union, November 27, 2001, https://www.aclu.org/other/use-facial-recognition-super-bowl-and-tampa. The technology is today hailed by many as being a way to increase security in banking, personal identification for unlocking devices, and other commercial uses.

29. The face recognition company Clearview AI in particular has been hit with numerous lawsuits over their data scraping. The history of the company is fascinating and has been the subject of a lot of journalistic coverage. The state of Illinois offers a great deal of protection against data scraping, so companies have been scrambling to remove all Illinois-based activity before lawsuits are filed there. Louise Matsakis, "Scraping the Web Is a Powerful Tool. Clearview AI Abused It," *Wired*, January 25, 2020, https://www.wired.com/story/clearview-ai-scraping-web/. See also Kashmir Hill, "Facial Recognition Start-Up Mounts a First Amendment Defense," *New York Times*, August 11, 2020, https://www.nytimes.com/2020/08/11/technology/clearview-floyd-abrams.html; Kashmir Hill, "Wrongfully Accused by an Algorithm," *New York Times*, June 24, 2020, https://www.nytimes.com/2020/06/24/technology/facial-recognition-arrest.html.

30. Perkowitz, "The Bias in the Machine."

31. Drew Harwell, "Wrongfully Arrested Man Sues Detroit Police over False Facial Recognition Match," *Washington Post*, April 13, 2021, https://www.washingtonpost.com/technology/2021/04/13/facial-recognition-false-arrest-lawsuit/.

32. Aaron Mak, "Amazon's Facial Recognition Tool Falsely Matched 28 Members of Congress to Mug Shots," *Slate*, July 26, 2018, https://slate.com/technology/2018/07/amazon-face-matching-technology-misidentified-28-members-of-congress-as-criminals.html.

33. W. Zhao et al., "Face Recognition: A Literature Survey," *ACM Computing Surveys* 35, no. 4 (December 1, 2003): 399–458, https://doi.org/10.1145/954339.954342; Agre, "Your Face Is Not a Bar Code."

34. Zhao et al., "Face Recognition," 399.

35. Zhao et al., 400.

36. Zhao et al., 453.

37. Zhao et al., 403.

38. Zhao et al., 453–54.

39. Zhao et al., 401.

40. Some recent work on this includes Yu Zhou et al., "The Constancy of the Holistic Processing of Unfamiliar Faces: Evidence from the Study-Test Consistency Effect and the within-Person Motion and Viewpoint Invariance," *Attention, Perception, & Psychophysics* 83, no. 5 (July 1, 2021): 2174–88, https://doi.org/10.3758/s13414-021-02255-8; Hoo Keat Wong, Alejandro J. Estudillo, Ian D. Stephen, and David R. T. Keeble, "The Other-Race Effect and Holistic Processing across Racial Groups," *Scientific Reports* 11, no. 1 (April 19, 2021): 8507, https://doi.org/10.1038/s41598-021-87933-1; Østergaard Knudsen, Katrine Winther Rasmussen, and Christian Gerlach, "Gender Differences in Face Recognition: The Role of Holistic Processing," *Visual Cognition* 29, no. 6 (July 3, 2021): 379–85, https://doi.org/10.1080/13506285.2021.1930312.

41. Brownlee, "A Gentle Introduction to Deep Learning for Face Recognition."

42. Arthur L. Caplan, "How Should We Regard Information Gathered in Nazi Experiments?," *AMA Journal of Ethics* 23, no. 1 (January 1, 2021): 55–58, https://doi.org/10.1001/amajethics.2021.55.

43. Lee Vinsel, "You're Doing It Wrong: Notes on Criticism and Technology Hype," Medium, February 1, 2021, https://sts-news.medium.com/youre-doing-it-wrong-notes-on-criticism-and-technology-hype-18b08b4307e5; Judy Wajcman, "Automation: Is It Really Different This Time?," *The British Journal of Sociology* 68, no. 1 (2017): 119–27, https://doi.org/10.1111/1468-4446.12239.

44. Vinsel, "You're Doing It Wrong."

45. Natasha Lomas, "Clearview AI in Hot Water Down Under," *TechCrunch* (blog), November 3, 2021, https://social.techcrunch.com/2021/11/03/clearview-ai-australia-privacy-breach/.

46. "740 ILCS 14/ Biometric Information Privacy Act," Illinois General Assembly, https://www.ilga.gov/legislation/ilcs/ilcs3.asp?ActID=3004&ChapterID=57; Tucker Guinn, "Illinois Court Denies Clearview AI's Motion to Dismiss Privacy Lawsuit," *Jurist*, August 31, 2021, https://www.jurist.org/news/2021/08/illinois-court-denies-clearview-ais-motion-to-dismiss-privacy-lawsuit/.

47. Tim Cushing, "Facial Recognition's Latest Failure Is Keeping People from Accessing Their Unemployment Benefits," *Techdirt*, June 29, 2021, https://www.techdirt.com/articles/20210620/16473147029/facial-recognitions-latest-failure-is-keeping-people-accessing-their-unemployment-benefits.shtml.

48. Corin Faife, "Feds Are Still Using ID.Me to Scan Your Face—and Human Reviewers Can't Keep Up," *The Verge*, February 11, 2022, https://www.theverge.com/2022/2/11/22928082/id-me-irs-facial-recognition-overworked-employees.

49. This is a significant growth area in medical technology that has been extensively covered. For a brief snapshot see Zhouxian Pan et al., "Clinical Application of an Automatic Facial Recognition System Based on Deep Learning for Diagnosis of Turner Syndrome," *Endocrine* 72, no. 3 (June 2021): 865–73, https://doi.org/10.1007/s12020-020-02539-3; Nicole Martinez-Martin, "What Are Important Ethical Implications of Using Facial Recognition Technology in Health Care?," *AMA Journal of Ethics* 21, no. 2 (February 1, 2019): 180–87,

https://doi.org/10.1001/amajethics.2019.180; Alexandra Reeve Givens, "How Algorithmic Bias Hurts People with Disabilities," *Slate*, February 6, 2020, https://slate.com/technology /2020/02/algorithmic-bias-people-with-disabilities.html.; Dian Hong et al., "Genetic Syndromes Screening by Facial Recognition Technology: VGG-16 Screening Model Construction and Evaluation," *Orphanet Journal of Rare Diseases* 16, no. 1 (August 3, 2021): 344, https://doi.org/10.1186/s13023-021-01979-y.

50. There are some brilliant Twitter threads discussing the challenges of doing ethical work in the domain of big data and AI; see, for example, the hashtag #IStandWithTimnit following the firing of Google ethics researcher Timnit Gebru.

51. For example, the Israel/Palestine border makes ample use of face recognition software and surveillance; for more on reactions to this use, see Olivia Solon, "Microsoft Funded Firm Doing Secret Israeli Surveillance on West Bank," NBC News, October 28, 2019, https:// www.nbcnews.com/news/all/why-did-microsoft-fund-israeli-firm-surveils-west-bank -palestinians-n1072116; Amitai Ziv, "This Israeli Face-Recognition Startup Is Secretly Tracking Palestinians," *Haaretz*, July 15, 2019, https://www.haaretz.com/israel-news/business /.premium-this-israeli-face-recognition-startup-is-secretly-tracking-palestinians-1 .7500359.

52. For more on the stakes for the poles of a spectrum, see Anne McGuire, *War on Autism: On the Cultural Logic of Normative Violence* (Ann Arbor: University of Michigan Press, 2016); Judith Butler, *Gender Trouble: Feminism and the Subversion of Identity*, 1st ed. (New York: Routledge, 2006).

53. For more on the role internet quizzes have played in rising awareness around face recognition, see chapter 2.

54. Sarah Bate et al., "Applied Screening Tests for the Detection of Superior Face Recognition," *Cognitive Research: Principles and Implications* 3, no. 1 (June 27, 2018): 22, https:// doi.org/10.1186/s41235-018-0116-5.

55. Just as someone bilingual is needed to translate, say, a massive amount of English to French for Google Translate to work, so too does FRT need someone to verify the comparisons between faces. Machine learning happens only because of human expertise. The history of these processes is one of building on human abilities and then effacing them, but let us not forget how deeply people are embedded into what the machines do. As Stephanie Dick so cogently has demonstrated, human cognition and machine cognition are different. We are not making the machines act like us. Machines cannot teach us how brains recognize faces, much as they cannot teach us how brains translate language. And while the outcomes look similar, and indeed the machines may develop expertise on a scale that people cannot achieve, we must create a framework for understanding what happens inside the impenetrable black box. Otherwise we may forget the role of people in creating that very box, and we may misrecognize its potentialities and indeed its dangers. There is no magic here. The machines can only do so much, and, without people, that's not much at all in the long term. They can't, for example, teach face-blind people how to recognize faces. Because brains work differently than machines.

56. Keyes, "The Bones We Leave Behind."

57. The story of the Algorithmic Justice League appears on its website: https://www.ajl .org/about.

Chapter 7. Is There Dyslexia without Reading?

1. Annemarie Jutel, *Diagnosis: Truths and Tales* (Toronto: University of Toronto Press, 2019); Steven G. Epstein, "The Politics of Health Mobilization in the United States: The Promise and Pitfalls of 'Disease Constituencies,'" *Social Science and Medicine* 165 (September 1, 2016): 246–54, https://doi.org/10.1016/j.socscimed.2016.01.048; Ian Hacking, *The Social Construction of What?* (Cambridge, MA: Harvard University Press, 1999); Rayna Rapp and Faye D. Ginsburg, "Enabling Disability: Rewriting Kinship, Reimagining Citizenship," *Public Culture* 13, no. 3 (September 1, 2001): 533–56.

2. Rajendra Prasad, "Historical Aspects of Milk Consumption in South, Southeast, and East Asia," *Asian Agri-History* 21, no. 4 (2017): 287–307, https://www.asianagrihistory.org/pdf /articles/Rajendra-Prasad-21-4.pdf; Fred Beauvais, "American Indians and Alcohol," *Alcohol Health and Research World* 22, no. 4 (1998): 253–59, https://www.ncbi.nlm.nih.gov/pmc /articles/PMC6761887/.

3. Hacking, *The Social Construction of What?*

4. For an outline of the neurology of prosopagnosia, J. A. Nunn, P. Postma, and R. Pearson, "Developmental Prosopagnosia: Should It Be Taken at Face Value?," *Neurocase* 7, no. 1 (January 1, 2001): 15–27, https://doi.org/10.1093/neucas/7.1.15.

5. We can spend time debating the differences between Darnton's cultural history and that practiced by scholars including Jacob Burkhardt, Johan Huizinga, and the Annales School. Darnton's scholarship represented a turn (back) to symbols and meaning following a long period of social history that focused on cause and structure. History and scholarly trends are cyclical, and they respond to both other trends and broader cultural and contextual factors, much like what we study.

6. I feel compelled to add here that reading is not a uniquely Western phenomenon, nor is non-literacy excluded from it. Jutel, *Diagnosis*, 3. Jutel does not return to the specific issue of dyslexia in the text.

7. Philip Kirby, "What's in a Name? 'Word Blindness' Was a Recognised Condition More than a Century Ago. But It Was Not until the 1970s That It Began to Be Accepted by the Medical Establishment," *History Today* 68, no. 2 (February 2018): 48–57, https://www.historytoday .com/archive/feature/history-dyslexia.

8. The sight metaphor in the Western tradition dates to antiquity; Heraclitus, for example, discusses eyes as witness, and Plato's allegory references visual knowledge and sunlight. In one of the most well-known manifestations of this metaphor, Descartes's sight phenomenology draws on the body's eye and the mind's eye.

9. Kirby, "What's in a Name?"

10. "The History of Nuts," Nutcracker Museum, http://www.nutcrackermuseum.com /history_nuts.htm.

11. Reading has a long history, with the first examples of written communication dating back to at least 3500 BCE. Books emerged much later. "A Brief History of Literacy," University of Texas, September 9, 2015, https://academicpartnerships.uta.edu/articles /education/brief-history-of-literacy.aspx.

12. Hacking, *The Social Construction of What?*, 58.

13. Ian Hacking, *Mad Travelers: Reflections on the Reality of Transient Mental Illnesses* (Charlottesville, University of Virginia Press, 1998).

14. Ian Hacking, "The Looping Effects of Human Kinds," in *Causal Cognition: A Multidisciplinary Debate*, 351–94, Symposia of the Fyssen Foundation (New York: Clarendon Press/Oxford University Press, 1995).

15. Joan Jacobs Brumberg, *The Body Project: An Intimate History of American Girls* (New York: Vintage, 1998), 138.

16. Susan E. Lederer, *Subjected to Science: Human Experimentation in America before the Second World War* (Baltimore: Johns Hopkins University Press, 1997); Harriet A. Washington, *Medical Apartheid: The Dark History of Medical Experimentation on Black Americans from Colonial Times to the Present* (New York: Anchor, 2008); Erika Dyck, *The Uses of Humans in Experiment* (Leiden: Brill, 2016); Kurt Danziger, *Constructing the Subject: Historical Origins of Psychological Research* (Cambridge: Cambridge University Press, 1994); Londa Schiebinger, *Secret Cures of Slaves: People, Plants, and Medicine in the Eighteenth-Century Atlantic World*, 1st ed. (Stanford, CA: Stanford University Press, 2017); Anita Guerrini, *Experimenting with Humans and Animals: From Galen to Animal Rights* (Baltimore: Johns Hopkins University Press, 2003).

17. Peter Conrad and Kristin K. Barker, "The Social Construction of Illness: Key Insights and Policy Implications," *Journal of Health and Social Behavior* 51, no. 1 suppl (March 1, 2010): S67–79, https://doi.org/10.1177/0022146510383495.

18. Charles E. Rosenberg and Janet Lynne Golden, *Framing Disease: Studies in Cultural History* (New Brunswick, NJ: Rutgers University Press, 1992); Charles E. Rosenberg, "Disease in History: Frames and Framers," *The Milbank Quarterly* 67 (1989): 1–15, https://doi.org/10.2307/3350182.

19. John C. Burnham, "Why Sociologists Abandoned the Sick Role Concept," *History of the Human Sciences* 27, no. 1 (February 1, 2014): 70–87, https://doi.org/10.1177/0952695113507572.

20. Barbara Ehrenreich and Deirdre English, *For Her Own Good: 150 Years of the Experts' Advice to Women* (London: Pluto Press, 1979).

21. Anna Cheshire et al., "Sick of the Sick Role: Narratives of What 'Recovery' Means to People With CFS/ME," *Qualitative Health Research* 31, no. 2 (January 1, 2021): 298–308, https://doi.org/10.1177/1049732320969395.

22. Tobin Siebers, *Disability Aesthetics* (Ann Arbor: University of Michigan Press, 2010); Rosemarie Garland-Thomson, *Extraordinary Bodies: Figuring Physical Disability in American Culture and Literature*, 20th ed. (New York: Columbia University Press, 2017); Rapp and Ginsburg, "Enabling Disability"; Dan Goodley, *Disability Studies: An Interdisciplinary Introduction* (Los Angeles: SAGE, 2011); Edmund Coleman-Fountain and Janice McLaughlin, "The Interactions of Disability and Impairment," *Social Theory & Health* 11, no. 2 (2013): 133–50, https://doi.org/10.1057/sth.2012.21; Lennard J. Davis, "Constructing Normalcy," in *The Disability Studies Reader*, edited by Lennard J. Davis, 3rd ed., 9–28 (New York: Routledge, 2010).

23. For example, wheelchair users have an impairment that makes it difficult or impossible to walk. The lack of curb cuts and wheelchair-accessible buildings is disabling. Making spaces wheelchair friendly would offer accommodations.

24. On March 13, 1990, over one thousand people marched from the White House to the Capitol Building to advocate for the passage of the Americans with Disabilities Act (ADA). When they reached the Capitol, some sixty protestors left their wheelchairs and mobility aids to climb up the steps of the Capitol to demonstrate the need for better accommodation and the urgency of the ADA. The "Capitol Crawl" is a key moment in disability advocacy and history, and it may well have contributed to the passage of the act on July 26, 1990.

25. Alison Kafer, *Feminist, Queer, Crip* (Bloomington: Indiana University Press, 2013).

26. A. J. Larner, "Retrospective Diagnosis: Pitfalls and Purposes," *Journal of Medical Biography* 27, no. 3 (August 1, 2019): 127–28, https://doi.org/10.1177/0967772019868433; Osamu Muramoto, "Retrospective Diagnosis of a Famous Historical Figure: Ontological, Epistemic, and Ethical Considerations," *Philosophy, Ethics, and Humanities in Medicine: PEHM* 9 (May 28, 2014): 10, https://doi.org/10.1186/1747-5341-9-10; Katherine Foxhall, "Making Modern Migraine Medieval: Men of Science, Hildegard of Bingen and the Life of a Retrospective Diagnosis," *Medical History* 58, no. 3 (July 2014): 354–74, https://doi.org/10.1017/mdh .2014.28.

27. Jason J. S. Barton and Sherryse Corrow, "The Problem of Being Bad at Faces," *Neuropsychologia* 89 (August 2016): 119–24, https://doi.org/10.1016/j.neuropsychologia.2016 .06.008.

28. Sharrona Pearl, "A Super Useless Super Hero," *Semiotic Review* 7 (September 20, 2019), https://www.semioticreview.com/ojs/index.php/sr/article/view/41.

29. Pearl.

30. Sergio Della Sala and Andrew W. Young, "Quaglino's 1867 Case of Prosopagnosia," *Cortex* 39, no. 3 (2003): 533–40, https://www.sciencedirect.com/science/article/abs/pii /S0010945208702636?via%3Dihub; A. L. Wigan, "The Duality of the Mind, Proved by the Structure, Functions, and Diseases of the Brain," *The Lancet* 43, no. 1074 (March 30, 1844): 39–41, https://doi.org/10.1016/S0140-6736(02)35039-6; John Hughlings Jackson, "Case of Disease of the Brain—Left Hemiplegia—Mental Affection," *Medical Times and Gazette*, May 4, 1872, 513–14; J. M. Charcot and D. Bernard, "Un cas de suppression brusque et isolée de la vision mentale des signes et des objets (formes et couleurs)," *Le Progrès Médical* 11 (1883): 568–71; Michigan State Medical Society, *Transactions of the Michigan State Medical Society* 23 (1899), https://www.google.com/books/edition/Transactions_of_the_Michigan _State_Medic/o97hAAAAMAAJ?hl=en&gbpv=1&dq=%22utterly+unable+to+distinguish+ some,+or+in+rare+instances,+all+colors.%22&pg=PA379&printsec=frontcover.

31. David R. Olson and Nancy Torrance, *The Cambridge Handbook of Literacy* (Cambridge: Cambridge University Press, 2009).

32. Hadyn D. Ellis and Melanie Florence, "Bodamer's (1947) Paper on Prosopagnosia," *Cognitive Neuropsychology* 7, no. 2 (March 1, 1990): 81–105, https://doi.org/10.1080 /02643299008253437.

33. Oliver Sacks, "Face-Blind," *New Yorker*, August 23, 2010, https://www.newyorker.com /magazine/2010/08/30/face-blind.

34. Pearl, "A Super Useless Super Hero."

35. Tom Campbell, "From Aphasia to Dyslexia, a Fragment of a Genealogy: An Analysis of the Formation of a 'Medical Diagnosis,'" *Health Sociology Review* 20, no. 4 (December 1, 2011): 450–61, https://doi.org/10.5172/hesr.2011.20.4.450.

36. Kirby, "What's in a Name?"

37. Sara Hendren, *What Can a Body Do? How We Meet the Built World* (New York: Riverhead, 2020); Rosemarie Garland-Thomson, "Misfits: A Feminist Materialist Disability Concept," *Hypatia* 26, no. 3 (2011): 591–609, https://doi.org/10.1111/j.1527-2001.2011.01206.x; Simi Linton, *Claiming Disability: Knowledge and Identity* (New York: New York University Press, 1998); Susan Wendell, *The Rejected Body: Feminist Philosophical Reflections on Disability*, 1st ed. (New York: Routledge, 2013).

38. Hendren, *What Can a Body Do?*; Thomson, *Extraordinary Bodies*; Linton, *Claiming Disability*.

39. A robust body of literature explores links between auditory processing and dyslexia. See, for example, Gerd Schulte-Körne and Jennifer Bruder, "Clinical Neurophysiology of Visual and Auditory Processing in Dyslexia: A Review," *Clinical Neurophysiology* 121, no. 11 (November 1, 2010): 1794–1809, https://doi.org/10.1016/j.clinph.2010.04.028.

40. The history of medical discrimination is long and horrifying, and the recent COVID-19 pandemic has thrown the implications of structural racism and poverty into sharp relief. Two classic pieces discussing the history of discriminatory medical experimentation include Allen M. Hornblum, *Acres of Skin: Human Experiments at Holmesburg Prison*, 1st ed. (New York: Routledge, 1999); Allan M. Brandt, "Racism and Research: The Case of the Tuskegee Syphilis Study," *The Hastings Center Report* 8, no. 6 (1978): 21–29, https://dash.harvard.edu/handle/1/3372911. There are many more. For two recent and powerful explorations of medical racism and the COVID-19 pandemic, see Roxane Gay, "Opinion | Remember, No One Is Coming to Save Us," *New York Times*, May 30, 2020, https://www.nytimes.com/2020/05/30/opinion/sunday/trump-george-floyd-coronavirus.html; Sabrina Strings, "Opinion | It's Not Obesity. It's Slavery," *New York Times*, May 25, 2020, https://www.nytimes.com/2020/05/25/opinion/coronavirus-race-obesity.html.

41. For a nuanced investigation into retrospective diagnoses and its limitations, see Randall Packard, "'Break-Bone' Fever in Philadelphia, 1780: Reflections on the History of Disease," *Bulletin of the History of Medicine* 90, no. 2 (June 1, 2016): 193–221.

42. Classic texts like Barbara Duden's *The Woman beneath the Skin* and Shigehisa Kuriyama's *The Expressiveness of the Body* have explored embodiment, medicine, the history of the body and the nature of the somatic in different contexts; reading those texts, in their own kind of looping effect, changes how the reader experiences her own body and her own corporality. More recent work like Sara Hendren's *What Can a Body Do?* reframed the limits of the body as possibilities, looking at prosthetics that draw on tacit and somatic knowledge to meet the specific needs of particular people and particular bodies in navigating the world. Barbara Duden, *The Woman beneath the Skin: A Doctor's Patients in Eighteenth-Century Germany*, rev. ed. (Cambridge, MA: Harvard University Press, 1998); Shigehisa Kuriyama, *The Expressiveness of the Body and the Divergence of Greek and Chinese Medicine* (New York: Zone Books, 2002); Hendren, *What Can a Body Do?*

43. Hacking, *The Social Construction of What?*

44. Many scholars have explored the indexical role of the face and its relationship to ethics and relationship. See, for example, Sharrona Pearl, *About Faces: Physiognomy in Nineteenth-Century Britain* (Cambridge, MA: Harvard University Press, 2010); Sharrona Pearl, *Face/On; Face Transplants and the Ethics of the Other*, 1st ed. (Chicago: University of Chicago Press, 2017); Richard Sennett, *The Fall of Public Man* (New York: W. W. Norton, 1992); Erving Goffman, *The Presentation of Self in Everyday Life*, 1st ed. (New York: Anchor, 1959); Gilles Deleuze and Felix Guattari, *A Thousand Plateaus: Capitalism and Schizophrenia*, trans. Brian Massumi, 1st ed. (Minneapolis: University of Minnesota Press, 1987); Emmanuel Levinas and Alphonso Lingis, *Totality and Infinity: An Essay on Exteriority* (Pittsburgh, PA: Duquesne University Press, 1969); Heather Laine Talley, *Saving Face: Disfigurement and the Politics of Appearance* (New York: New York University Press, 2014).

45. The meaning of noses in particular has changed significantly over time. Another powerful example of changing visual signifiers can be found in the Victorian representation of the Irish in both England and the United States. Pearl, *About Faces*, 106–47; Sharrona Pearl, "White, with a Class-Based Blight: Drawing Irish Americans," *Éire-Ireland* 44, no. 3–4 (2010): 171–99, https://doi.org/10.1353/eir.0.0045.

46. As John Durham Peters pointed out to me, there can in fact be medical implications to eye color.

47. Equally, the myth of the uniquely Jewish big nose is, in fact, just a myth. Sharrona Pearl, "The Myth of the Jewish Nose," *Tablet* magazine, February 8, 2019, /sections /community/articles/the-myth-of-the-jewish-nose.

48. Jia Tolentino, "The Age of Instagram Face," *New Yorker*, December 12, 2019, https:// www.newyorker.com/culture/decade-in-review/the-age-of-instagram-face.

49. John Durham Peters, *The Marvelous Clouds: Toward a Philosophy of Elemental Media* (Chicago: University of Chicago Press, 2016), 22.

50. Peters, 89.

51. Peters, 35.

52. Peters, 22.

53. Allan Sekula, "The Body and the Archive," *October* 39 (1986): 3–64, https://www.jstor .org/stable/778312 .

54. Zohar Kampf, "Journalists as Actors in Social Dramas of Apology," *Journalism* 12, no. 1 (January 1, 2011): 71–87, https://doi.org/10.1177/1464884910385190.

55. Lisa Parks, "Technostruggles and the Satellite Dish: A Populist Approach to Infrastructure," in *Cultural Technologies: The Shaping of Culture in Media and Society*, ed. Göran Bolin (New York: Routledge, 2012), 64–84, https://doi.org/10.4324/9780203117354-11; Geoffrey C. Bowker and Susan Leigh Star, *Sorting Things Out: Classification and Its Consequences*, rev. ed. (Cambridge, MA: MIT Press, 2000); Peters, *The Marvelous Clouds*, 35–38.

56. Sybille Krämer, *Medium, Messenger, Transmission: An Approach to Media Philosophy*, o ed. (Amsterdam: Amsterdam University Press, 2015).

57. People in persistent vegetative state (PVS) may have functioning bodies but their status is highly contested. Kurt Gray, T. Anne Knickman, and Daniel M. Wegner, "More Dead than Dead: Perceptions of Persons in the Persistent Vegetative State," *Cognition* 121, no. 2 (November 2011): 275–80.

58. Bruno Latour, *Reassembling the Social: An Introduction to Actor-Network-Theory*, 1st ed. (Oxford: Oxford University Press, 2007).

59. Deleuze and Guattari, *A Thousand Plateaus*; Martin Müller, "Assemblages and Actor-Networks: Rethinking Socio-Material Power, Politics and Space," *Geography Compass* 9, no. 1 (2015): 27–41, https://doi.org/10.1111/gec3.12192.

60. Rachel Colls, "Feminism, Bodily Difference and Non-Representational Geographies," *Transactions of the Institute of British Geographers* 37, no. 3 (2012): 430–45; Rachel Slocum, "Thinking Race through Corporeal Feminist Theory: Divisions and Intimacies at the Minneapolis Farmers' Market," *Social & Cultural Geography* 9, no. 8 (December 1, 2008): 849–69, https://doi.org/10.1080/14649360802441465; Jasbir K. Puar, *Terrorist Assemblages: Homonationalism in Queer Times* (Durham, NC: Duke University Press, 2017).

61. Bowker and Star, *Sorting Things Out*.

62. Campbell, "From Aphasia to Dyslexia, a Fragment of a Genealogy"; Kirby, "What's in a Name?"; P. Kirby, "Worried Mothers? Gender, Class and the Origins of the 'Dyslexia Myth,'" *Oral History* 47, no. 1 (April 1, 2019), https://ora.ox.ac.uk/objects/uuid:6e177352-6dc7 -4c34-8d90-4fcb75528871.

63. Philip Kirby, "Gift from the Gods? Dyslexia, Popular Culture and the Ethics of Representation," *Disability & Society* 34, no. 9–10 (November 26, 2019): 1573–94, https://doi.org /10.1080/09687599.2019.1584091.

64. Deleuze and Guattari, *A Thousand Plateaus*.

65. For a powerful personal testimony that heavily supports the link between genius and mental illness and neurodiversity, see Kay Redfield Jamison, *An Unquiet Mind: A Memoir of Moods and Madness*, 1st ed. (New York: Vintage, 1996).

66. Kirby, "Gift from the Gods?"

67. See, for example, Avinoam B. Safran and Nicolae Sanda, "Color Synesthesia. Insight into Perception, Emotion, and Consciousness," *Current Opinion in Neurology* 28, no. 1 (February 2015): 36–44, https://doi.org/10.1097/WCO.0000000000000169.

68. Octavia E. Butler, *Lilith's Brood: The Complete Xenogenesis Trilogy* (New York: Open Road Media Sci-Fi & Fantasy, 2012).

69. This is a key component of Alison Kafer's work. Kafer, *Feminist, Queer, Crip*.

70. Riva Lehrer, *Golem Girl: A Memoir* (New York: One World, 2020).

71. Here I follow Lehrer's usage of capitalizing "Disability" in the context of Disability culture and community, and otherwise using the lower case.

72. Tali Bitan et al., "Morphological Decomposition Compensates for Imperfections in Phonological Decoding. Neural Evidence from Typical and Dyslexic Readers of an Opaque Orthography," *Cortex* 130 (September 1, 2020): 172–91, https://doi.org/10.1016/j.cortex.2020 .05.014.

73. See, for example, the writing of Emmeline May about her adaptations, discussed in Pearl, "Watching While (Face) Blind: Clone Layering and Prosopagnosia," 87–88.

74. Michael A. Webster, "Color Vision: Glasses Half Full," *Current Biology* 30, no. 16 (August 17, 2020): R952–54, https://doi.org/10.1016/j.cub.2020.06.062.

Conclusion. Beyond the Face

1. Sharrona Pearl, "Watching While (Face) Blind: Clone Layering and Prosopagnosia," in *Orphan Black: Performance, Gender, Biopolitics*, ed. Andrea Goulet and Robert A. Rushing (London: Intellect Ltd., 2019), 78–91, http://www.press.uchicago.edu/ucp/books/book /distributed/O/bo31275041.html.

Coda. The Detective Story

1. Thucydides, *History of the Peloponnesian War* 2.

2. There may well be other historical examples from other literatures and countries with which I am not familiar and that I did not discover in my research; if so, I would be eager to learn about them.

Index

Milton Keynes UK
Ingram Content Group UK Ltd.
UKHW040607031123
431765UK00001B/1